Electrical Properties of Rocks

Monographs in Geoscience
General Editor: Rhodes W. Fairbridge
Department of Geology, Columbia University, New York City

B. B. Zvyagin
  Electron-Diffraction Analysis of Clay Mineral Structures–1967

E. I. Parkhomenko
  Electrical Properties of Rocks–1967

In preparation

A. I. Perel'man
  The Geochemistry of Epigenesis

Electrical Properties of Rocks

E. I. Parkhomenko
Institute of Physics of the Earth
Academy of Sciences of the USSR, Moscow

Translated from Russian and edited by
George V. Keller
Colorado School of Mines
Golden, Colorado

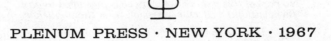

PLENUM PRESS · NEW YORK · 1967

Eleonora Ivanovna Parkhomenko, a senior scientist in the department of physical properties of rocks at the Institute of Physics of the Earth of the Academy of Sciences of the USSR in Moscow, graduated from the Mining Institute in Irkutsk in 1943. From 1943 to 1950 she worked in an industrial laboratory and joined the Institute of Physics of the Earth in 1950. In 1957, she was awarded the degree of Candidate in the Physical and Mathematical Sciences. She has published 34 papers and at present she is studying the electrical properties of rocks at high pressures and temperatures.

The Russian text has been corrected by the author for the present edition. It was originally published for the O. Yu. Shmidt Institute of Physics of the Earth of the Academy of Sciences of the USSR by Nauka Press in Moscow under the following title:

Электрические свойства горных пород

ELEKTRICHESKIE SVOISTVA GORNYKH POROD

ELECTRICAL PROPERTIES OF ROCKS

Э. И. Пархоменко

Library of Congress Catalog Card Number 67-10311

Foreword

Recently there has been growing interest in the physical properties of rocks. To interpret data on the geophysical fields observed near the Earth's surface, we must know the physical properties of the materials composing the interior. Moreover, the development of geophysical methods (in particular, electrical methods) is necessitating a multiple approach to the study of the physical properties of rocks and minerals.

In connection with problems now appearing, the physical properties of rocks must be studied in the laboratory under various thermodynamic conditions. Electrical methods of geophysical exploration often may require only data obtained at atmospheric pressure and room temperature, or at temperatures below 100°C. If, however, we have in mind geophysical field observations on the composition and state of matter deep in the Earth's crust and mantle, we must conduct laboratory experiments at high pressures and temperatures. For example, in interpreting data from geomagnetic soundings of the mantle, we may need experimental results on the electrical properties of rocks at pressures of tens of kilobars and temperatures of the order of 1000°C. In this connection, we must remember that pressure has relatively little effect on the electrical properties of rocks, whereas, temperature affects them very strongly.

At present, while research into the mechanical properties of rocks (relating to the problems of geophysics, geochemistry, geology, and mining) is pressing forward on a wide front, much less work is being done with electrical properties. Nevertheless, the electrical characteristics of rocks and minerals are of interest for a number of practical purposes.

Until recently, the electrical properties of rocks were studied mainly in connection with the requirements of well logging. The resultant data, however, have much wider uses in geophysics. In addition, the electrical properties of rocks are now being used in mining; there are promising prospects for the use of electrophysical rock-breaking methods in mining. For this purpose, we shall need to know how the electrical properties of rocks and minerals vary with various factors— in particular, temperature and pressure.

In recent years there have been extensive developments of various types of electrical prospecting, including electrical sounding and electrical profiling with direct current, the method of telluric currents, the induction, radio-wave transmission and electromagnetic frequency-sounding methods, and the piezoelectric method for prospecting for quartz and pegmatite veins. The use of these methods demands a knowledge of various characteristics of sedimentary, metamorphic, and igneous rocks. É. I. Parkhomenko's book therefore contains chapters giving detailed discussions of the methods and results of investigations of electrical resistivity, dielectric constant, and dielectric loss in rocks and minerals.

Rocks and rock-forming minerals are quite varied and complex in their electrical properties. For example, their resistivities vary from 10^{-2} to 10^{16} ohm $\cdot$ cm. Most of them are dielectrics, but some are semiconductors. Their electrical properties depend on their chemical and mineral contents, genesis and petrographic characteristics, structure, texture, porosity, water content, etc. The first chapter therefore gives a short summary of the petrography of rocks. In addition, later chapters include sections on the electrical properties of dielectrics and semiconductors.

This is the first monograph ever devoted specifically to the electrical properties of rocks. Owing to the increasing practical uses now being made of the electrical properties of rocks in electrical methods for prospecting for minerals, as well as for other purposes in geophysics, geology and mining, this monograph is a timely production. É. I. Parkhomenko's systematization of the accumulated material will undoubtedly be beneficial to the further development of the subject.

In conclusion, the reader will find descriptions of the piezoelectric and seismo-electric effects and certain other electrical phenomena observed in rocks, including high-voltage and induced polarization.

Mikhail P. Volarovich
Director, Department of the
Physical Properties of Rocks
Institute of Physics of the Earth
Academy of Sciences of the USSR

Contents

ELECTRICAL PROPERTIES

OF ROCKS

Chapter I

Brief Review of the Petrography of Rocks

The physical properties of a material cannot be studied independently of the structure and chemical composition of the material. Rocks commonly are very complex with respect to textural and chemical properties, consisting of multicomponent aggregates. The texture and composition of aggregates are of primary importance in determining physical properties. Therefore, before considering the reported values for electrical properties of rocks, it is necessary to review briefly the petrographic properties of various rocks. References [1-5] provide the best review of the general subject of petrography.

Mineral Composition of Rocks

Rocks are aggregates of mineral grains bound together by molecular interaction forces. Minerals which are chemical compounds and pure elements differ from one another in chemical composition, internal structure, and physical properties. The properties characterizing a mineral are: crystal form and grain habit, hardness, durability, ductility, cleavage, color, luster, and other secondary properties. The great majority of minerals have a crystal structure; that is, they have some orderly internal molecular structure. Some minerals occur in an amorphous form, as, for example, various forms of chert, volcanic glass, and so on.

1

At the present time, about 2000 minerals have been classified. Most of these minerals occur but rarely in nature, and only
a few tens of these minerals are common rock constituents. These
minerals determine the rock type and the rock name that they form;
for example, the compounds of granite are feldspar, muscovite,
and quartz.

It should be noted that one particular mineral may be essential in one rock type, and occur only as an accessory in another
rock type. For example, quartz is an essential mineral in granite,
but is an accessory mineral in gabbro. Sometimes, classification
is made also on the basis of minor minerals, as, for example,
fluorspar in granite or gneiss, galena in sandstone, sphalerite in
limestone, topaz in liparite, and so on.

The classification of minerals is based on chemical composition.

Feldspar is the commonest rock-forming mineral group.
Minerals in this group comprise about 60% of the earth's crust.
The total number of feldspar minerals with different compositions
is nearly 800. These are found largely in granite, syenite, diorite,
gabbro, basalt, diabase, and so on.

After the aluminosilicates, quartz is the commonest mineral
found in the earth's crust. It is the main constituent of quartzite
and sandstone. Quartz is also found in large quantities in igneous
rocks such as quartz porphyry, granite, granite porphyry, liparite,
and so on, as well as in the metamorphic rock, gneiss. Quartz
frequently occurs in fine-grained or coarsely crystalline form in
veins penetrating other rock types. Iron magnesium silicates and
carbonates are less abundant.

Rock may be classified as monomineralic (marble, labradorite, limestone) or as polymineralic (granite, gabbro, peridotite)
depending on whether it contains one or more essential minerals.
It should be noted that a monomineralic rock usually contains other
minerals as accessories.

The mechanical properties of a rock are determined to a
large degree by the properties of the minerals comprising the
rock. However, the electrical and magnetic properties in a number
of cases depend on the properties of accessory minerals; for
example, the magnetic properties of basalt, diabase, and granite

are determined by the presence of such minerals as magnetite, hematite, and pyrrhotite.

Basic Rock Types

Depending on their origin, rocks may be classified as igneous, sedimentary, or metamorphic.

The first type of rock is formed by the cooling of magma, or molten rock. If the cooling of the magma takes place slowly at great depths, the igneous rock is intrusive and has a coarsely crystalline form. Rocks solidifying at the surface or at shallow depths are termed extrusive. These typically are microcrystalline. Typical examples of such rocks are basalt, trachyte, andesite, and liparite. With very rapid cooling, the texture may be cryptocrystalline or amorphous, as in volcanic rocks.

Igneous rocks at the earth's surface break down as a result of various weathering processes (mechanical, thermal, and chemical) into fragments which differ in chemical composition. The mechanical deposition of these fragments leads to the formation of sedimentary rocks (sandstones). Sedimentary rocks may also be formed by precipitation of salts from solution (rock salt, gypsum). Such rocks are termed evaporites. Finally, shell fragments and debris from dead organisms accumulate into rock masses which are termed organic rocks (organic limestone, coal, oil).

The surface layers of the earth's crust usually are sedimentary rocks. The most common sedimentary rocks are shales, followed by sandstones and limestones.

Sedimentary rocks and extrusive igneous rocks may be submerged to great depths in the earth's crust by tectonic forces, where the temperature and pressure radically alter the properties of the rocks. Such altered rocks are termed metamorphic rocks. Intrusive igneous rocks may also be metamorphosed.

Depending on the degree of metamorphism, metamorphic rocks may be classified in the following series: gneiss, slate, quartzite, and marble.

All of these classes of rocks occur in the earth's crust, which extends to an average depth of 35 to 40 km under the continents, and to an average depth of 10 to 13 km under the oceans.

The greatest thickness of the crust in any region is 70 km. Usually the sequence of rock classes through progressively larger depths in the crust is sedimentary, metamorphic, igneous. It should be mentioned that at greater depths, basic and ultrabasic rocks become more important.

In addition to the classification of rocks on the basis of genesis, rocks are classified according to chemical and mineralogic composition, particularly the silica content. In such a classification, rocks are grouped as silicic, intermediate, basic, ultrabasic, and alkaline. The silicic rocks are rich in silica, usually occuring as quartz, and are represented by such types as granite, granodiorite, liparite, and quartz porphyry. In most rocks, all of the silica is found in combination (diorite, porphyrite, andesite, syenite-porphyry, and trachyte). Basic rocks are characterized by relatively small silica content. Examples of such rocks are gabbro, basalt, and pyroxenite. They contain 35 to 40% silica in chemical combination. The alkaline rocks include nepheline syenite, urthite, and others. Rocks in this group are characterized by a high content of alkaline metals.

Rock Structure

The basic structural characteristics of rocks are: (1) the degree of crystallization or crystallinity; (2) the size of individual crystals; (3) the grain habit; and (4) the interrelation between crystalline and glassy material.

The degree of crystallinity of rock is determined by the conditions under which it crystallized and the viscosity of the melt. If crystallization took place slowly at great depth, a rock consists entirely of well-formed mineral crystals. It has a crystalline grain structure. This type of structure is characteristic of granite, syenite, diorite, gabbro, peridotite, and so on. When a rock solidifies very quickly, it consists of glass (basalt glass, obsidian). Different varieties of basalt provide examples of a rock composed entirely of crystals or glass.

The relative size of individual grains defines two basic types of structures — microcrystalline and macrocrystalline. Grain diameters in a microcrystalline rock range from 0.001 to 0.1 mm. Macrocrystalline rocks are further divided into the following classes: fine-grained, medium-grained, large-grained, and coarse-grained. A fine-grained rock is considered to be one in which the grain size is less than 1 mm. If the grain size falls in the range 1 to 5 mm, the rock is termed medium-grained. A large-grained rock has grain sizes in the range 5 to 10 mm. Coarse-grained rocks have still larger grain diameters. Depending on whether or not the grain sizes are all about the same, a rock may be termed uniformly grained or porphyrytic. A uniformly grained rock is one in which the principal minerals all have about the same grain size.

When there are several distinctly different grain sizes in a rock, it is termed nonuniformly grained. There are two possible variants of such rocks, depending on the relationship between grains of various sizes. The larger crystals may be imbedded in a ground mass consisting of medium- or fine-grained rock or glass. This type of structure is called a porphyry. The opposite to a porphyry structure is a poikilitic structure, in which the finer-grained minerals are distributed between the larger-grained minerals.

Grain sizes and their interrelations have a significant effect on the physical properties of a rock and, therefore, they are a very important structural characteristic.

Another characteristic of structure which should be considered is the crystal form, which is determined by the dimensions of a grain in three directions. An isometric crystal is one which has about equal dimensions in three orthogonal directions. Crystals which are well-developed in only two directions may be platy or flat-faced (mica or some forms of feldspar). Crystals which develop principally along one direction are termed prismatic (amphibole or apatite) or needlelike.

With any of these structures, the individual grains may be arranged randomly and chaotically, so that the resultant rock appears to be isotropic.

Rock Texture

In addition to the rocks which lack any orderly arrangement of mineral grains, there are many rocks in which the grains exhibit some preferred orientation. This orientation of grains constitutes the texture of a rock.

Depending on the conditions of formation of a rock, texture may be considered to be primary or secondary. Primary texture is found in igneous rocks in which flow of the parent magma took place after one mineral had crystallized out of the melt. Flow of the suspension of such crystals would result in a preferred orientation in the final rock.

Primary texture in an igneous rock may be linear or planar. A linear texture is characterized by parallel orientation of prismatic and tabular minerals. A planar texture in an igneous rock is represented by alternating layers with different properties and by the plane-parallel distribution of tabular minerals. Primary layering in a sedimentary rock is represented by the orientation of grains in the bedding plane, by uniform mineral composition within a bed, or by alternating layers of different minerals.

Secondary texture in a rock arises from tectonic processes. It is characteristic of such metamorphic rocks as gneiss, quartzite, marble, and so on.

In considering texture caused by an orderly orientation of grains, two types of order may be recognized − orientation by the external shape of a crystal, or orientation by the internal crystal structure. Orientation by shape is possible when grains are elongate or tabular crystal forms. Such orientation may be either primary or secondary, resulting from grain deformation. On the other hand, frequently in quartzite, the quartz occurs as grains with isometric form but shows a preferential orientation in terms of internal crystal structure, that is, in terms of the axes of crystallization.

In some cases, grains may be oriented with respect to external shape and internal structure at the same time. Orientation with respect to external shape is usually observed visually, while orientation in terms of internal structure is done with a microscope, using special techniques developed by Zander and Federov for observing the optical axes of mineral grains in thin section.

Texture may be complete or limited, depending on the degree or orientation of the minerals. However, primarily there may be a variability in texture caused by the orientation or elements in a nonhomogeneous manner. The less this variability, the more pronounced will be the anisotropy of a rock caused by texture.

However, we should note that this anisotropy may not be observed in all physical properties; grains may be isotropic with respect to some physical properties and yet be anisotropic with respect to other physical properties.

Porosity and Water Content of Rocks

Pores and fractures in rocks are very important forms of nonuniformity of structure and texture.

Porosity is taken to be the total void space which exists between the solid mineral particles in a rock; these total void spaces arise from various causes and differ in shape and size. Pore sizes vary widely, ranging from gross cavities easily seen by the naked eye to micropores which have dimensions comparable to the size of a single molecule. Moreover, pores may be open, closed, or isolated. Generally, the porosity of a rock is taken to be the total volume of open and closed pores. Porosity is defined as the ratio of the volume of pores to the total volume of the rock. This porosity coefficient may be written as K_p for the total porosity, K_{po} for open porosity, and K_{pe} for effective porosity. The effective porosity is defined as the ratio of the volume of interconnected pores to the total volume of the rock.

Representative values for the porosity of various rock types are given in Table 1. We see from this that, depending on the type of cement and the depth of burial, the largest porosities are observed in sedimentary rocks. Commonly, porosity decreases with depth as a result of overburden pressure. As one would expect, intrusive igneous rocks have a porosity of, at most, 1 to 2%.

Porosity values are most important, not only in interpreting electric logs, but also in the study of the physical properties of rocks. At present, the opinion is held that the greatest pressure under which pores may stay open is of the order of 1000 to 1500 kg/cm^2 [9, 10].

The existence of pores and fractures in a rock results in the rock containing some quantity of water, which may be bound or free. The bound water in a rock may be closely bound or loosely bound. Closely bound water consists of oriented water molecules, adsorbed on the grain surfaces. Adsorbed water has rather special properties – an increased density, a low freezing point (− 78°C), a

Table 1. Porosity of Rocks

Rock type	Usual range of porosity, %		Reference
	from	to	
Shale	20	50	[6]
Sand	20	35	[6]
Sandstone	5	30	[6]
Limestone	1.5	15.0	[6]
Dolomite	3	20	[6]
Quartzite		0.8	[7]
Granite		1.2	[7]
Diabase		1.0	[7]
Biotite granite	0.4	5.2	[8]
Nepheline syenite	0.7	5.0	[8]
Diorite	1.3	5.1	[8]
Gabbro	0.3	3.5	[8]
Norite gabbro	1.3	2.0	[8]

decreased capability for dissolving salts, and so on. With weakly bound water, these anomalous properties differ to a lesser degree from those of free water. Variations in the amount of bound water lead to increases or decreases in the volume of water in a rock. Free water, moving under the effect of gravity, causes solution or precipitation of the solid constituents of a rock.

The water content of a rock may be expressed quantitatively in terms of the volume or weight of water or as water saturation. The volume fraction of water is expressed as the ratio of the volume of water filling the pores to the total volume of the rock. Water saturation is the ratio of the volume of water in the pore structure to the volume of the pores. Also, in the literature, the ratio of the weight of the water to the weight of the dry rock is frequently used as a measure of the moisture content.

Igneous and metamorphic rocks are characterized by low water content (less than 1%). The water content of sedimentary rocks varies over broad limits, depending on grain size and rock density.

In the case of oil reservoir rocks, information about permeability is most important [11] as is information about water content and porosity. This property is a measure of the ease with which liquid or gas may be produced from a rock. The permeability may determine the degree of saturation of a rock with oil, connate water, and so on, which in turn has an important effect on the physical properties of the rock.

Rocks vary widely in permeability. Friable and weakly cemented sands have high permeabilities. The coarser the grain size of the sand, the greater will be the permeability. Shales, siltstones, marls, and shaly sandstones with a high proportion of clayey cement have pores of subcapillary size and are practically impermeable.

In this brief discussion, I have attempted only to review those petrographic properties which are important in determining the physical properties of a rock.

Chapter II

Dielectric Properties of Rocks

The electrical properties of a material may be characterized in terms of a dielectric permeability, a resistivity and a dielectric loss. Dielectric permeability is the most important electrical property of rock which has a high resistivity. The study of this property is particularly important in relation to the use of the alternating-current methods of electrical exploration. Rocks are multiphase systems which consist of crystals, as well as amorphous solids, liquids, and gases. This complicates the study of physical properties. While the main constituents of a rock are the solid minerals, properties such as the dielectric permeability and electrical conductivity are determined mainly by the water content.

The dielectric properties of rocks have not been thoroughly studied. In the literature, many important questions have been raised concerning the effect of the dielectric permeability of the mineral framework, the moisture content, the frequency, the temperature, etc., on the bulk dielectric properties of a rock. Each of these areas requires further work. Before going ahead with a discussion of these properties, it will be useful to consider the basic theory of dielectric polarization, as is given in such monographs as references [12] and [13].

Dielectric Polarization

Dielectrics may consist of atoms, molecules, and ions which have the property of being unable to provide free conduction electrons (as in metals). Thus, a particular dielectric material may be characterized by limited movement of charged particles or orientation of polar molecules when an external electric field is applied. This motion of charged particles or rotation of polar molecules is termed dielectric polarization.

The fundamental characteristic of electric polarization in a dielectric material is the magnitude of the electric moment in a unit volume, $\bar{I}$, for the vector representing the intensity of polarization. It is equal to the product of the average dipole moment, m, for each particle and the number of polarized particles, n, per unit volume. The vector representing the intensity of polarization inside the material is directed from negative to positive charge.

In an isotropic dielectric, the intensity vector, $\bar{I}$, lies in the same direction as the vector for the applied electric field, $\bar{E}$, and the two are linearly related; $\bar{I} = \varkappa E$. The coefficient $\varkappa$ is termed the dielectric susceptibility.

In an anisotropic dielectric, the magnitude of the intensity vector depends on the direction of the electric field vector relative to the crystal axes. Therefore, it is necessary to write the relation between $\bar{I}$ and $\bar{E}$ in tensor form, which in the most general case would be

$$I_x = \varkappa_{11}E_x + \varkappa_{12}E_y + \varkappa_{13}E_z,$$
$$I_y = \varkappa_{21}E_x + \varkappa_{22}E_y + \varkappa_{23}E_z,$$
$$I_z = \varkappa_{31}E_x + \varkappa_{32}E_y + \varkappa_{33}E_z.$$

The values for the dielectric susceptibility $\varkappa_{ik}$ depend on the orientation of the xyz coordinate system with respect to the crystal axes. Because the susceptibility tensor is symmetric, there is some set of coodinates xyz for which the off-diagonal coefficients $\varkappa_{ik}$ are zero. These are termed the principle coefficients, and are designated as $\varkappa_1$, $\varkappa_2$, and $\varkappa_3$.

In practice, in investigating the dielectric properties of a material, the dielectric permeability is used rather than the dielectric susceptibility. The dielectric permeability, ε, is the ratio of dielectric displacement, D, to the voltage contributed by the electric field, E. As is well known, $\varepsilon = 1 + 4\pi\,\varkappa$, where $\varkappa$ is the dielectric susceptibility, so that $\varepsilon > 1$.

The dielectric permeability, like the dielectric susceptibility, is a scalar quantity for an isotropic medium, and is a tensor in the case of an anisotropic medium, with the coefficients depending upon the crystal directions.

The relationship between D and E in an anisotropic medium may be written in the following general form:

$$D_1 = \varepsilon_{11}E_x + \varepsilon_{12}E_y + \varepsilon_{13}E_z,$$
$$D_2 = \varepsilon_{21}E_x + \varepsilon_{22}E_y + \varepsilon_{32}E_z,$$
$$D_3 = \varepsilon_{31}E_x + \varepsilon_{32}E_y + \varepsilon_{33}E_z.$$

This tensor also is symmetric, and with the proper choice of the coordinate system, the off-diagonal terms will be zero, and the principal values may be designated as ε_1, ε_2, ε_3.

The electric polarization of a dielectric is a complicated physical phenomenon. The mechanics of polarization for dielectrics with different structures varies, depending on the presence of various types of polarization. Below, we will consider the factors which are most important.

Fundamental Types of Dielectric Polarization. Dielectrics may be divided into two classes – dielectrics in which the dipole moments of the elementary particles (atoms, molecules) or of the elements of the crystal lattice in an ionic crystal are zero, and dielectrics formed from polar molecules with some constant dipole moment. There are two corresponding types of polarization: (1) that due to motion of charged particles and (2) relaxation polarization. The latter occurs not only in materials formed of polar molecules, but also in material containing weakly bound ions. In addition, resonance polarization, observed only at high frequencies, is another basic form of polarization.

Polarization by electron motion results from movement of electrons relative to the nucleus in an atom or ion under the influence of an external electric field. As the electrons move away from the nuclei, the distance of travel is limited by a restoring force which develops between the electron and the nucleus. The less this restoring force is, the greater is the electrical dipole moment which will be developed by an external electric field.

Therefore, as this force diminishes, as is the case when an electron orbit is increased, the polarizability, α, of an atom increases. The polarizability of an atom is defined as being the coefficient of proportionality between the electric moment, $\overline{m}$, and the electric field, $\overline{E}$, applied at the center of the dipole, or $\overline{m} = \alpha\overline{E}$. Moreover, the polarizability of an atom increases with an increase in the number of electrons orbiting about it. Computations indicate that the polarizability of a particle equals the cube of the orbital radius. Results of measurements are in good agreement with such computations. The time required for polarization by electron displacement to take place is 10^{-15} sec, which is considerably less than the period of electromagnetic waves used in radio communications. As a result, the dielectric permeability of minerals caused by electron displacement shows no dispersion over the whole frequency range from zero to optical frequencies.

The dielectric permeability caused by electron displacement is independent of temperature. In this respect, in materials exhibiting only electron polarization, polarization decreases with increasing temperature. This decrease, which is readily seen in gases and nonpolar liquids, is caused by a decrease in density in the material with a corresponding decrease in the number of atoms per unit volume. Similarly, pressure has the opposite effect. Increasing pressure increases the density of a material and, therefore, also the dielectric permeability.

Polarization by electron displacement is the most widely observed type of polarization, being found in all materials, solid, liquid, or gaseous.

In addition to electron polarization, polarization may also occur because of ion displacement. In such a case, an applied electric field causes ions of one sign to be displaced relative to ions of the opposite sign. A force of Coulomb attraction exists between unlike ions in such molecules, and a force of repulsion exists between electrons in the surrounding electron cloud. In the absence of an external field, these interaction forces add to zero. An electric field distorts the balance condition, resulting in the displacement of ions relative to normal positions. The magnitude of the displacement and, therefore of the polarizability is larger for the larger ionic radii, inasmuch as the interaction forces diminish with increased distance between ion centers.

A rough computation of the polarizability caused by ion displacement indicates that it varies as the cube of the ionic radii. In cases in which ion displacement polarization is observed in addition to electron displacement polarization, increasing temperature causes the following effects. On one hand, the increasing temperature causes a decrease in the density of the material and, on the other hand, a weakening of the interionic bonding forces. Therefore, if the second effect is more important than the first, an increase in temperature causes an increase in ε. Experiments indicate that the dielectric permeability of cubic ionic crystals increases linearly, though insignificantly.

Polarization by ion displacement is observed in amorphous materials as well as in crystalline materials. Dielectric permeabilities for such materials fall in the range of 4 to 15. Polarization by ion displacement requires a time of 10^{-13} to 10^{-12} sec.

For materials exhibiting displacement type polarizations, the relation between polarizability and dielectric permeability is given by the Clausius—Mosotti equation

$$\frac{\varepsilon - 1}{\varepsilon + 2} = \frac{4\pi}{3} n\alpha, \qquad\qquad (II.1)$$

where n is the number of particles per unit volume and α is the polarizability.

Equation (II.1a) may be derived from equation (II.1) by considering the difference between the macroscopic electric field and the electric field actually present in the immediate vicinity of the elementary dipoles. A difference is found in any material with a dielectric permeability greater than unity which exhibits significant polarization. For dielectrics in which polarization is contributed solely by electron displacement, the dielectric permeability is very nearly the equal of the index of refraction, according to Maxwell's equation $\varepsilon = \nu^2$. Substituting this in equation (11.1), we have

$$\frac{\nu^2 - 1}{\nu^2 + 2} = \frac{4\pi}{3} n\alpha.$$

The Clausius—Mosotti relation is valid only for materials which polarize weakly and which do not contain polar molecules, radicals, and so on.

Relaxation polarization takes place in dielectrics containing polar molecules, or molecules with polar radicals, and weakly bonded ions, as well as in materials in which electron defects are developed by thermal activation. Such materials exhibit various types of relaxation polarization — dipole, ion, and electron. In general, for all these forms of polarization there is a close relationship with the thermal mobility of particles which exists independently of other factors. Materials exhibiting this type of polarization have high values for dielectric permeability.

Dipole relaxation polarization in dielectrics consists of the orientation of dipoles under an applied electric field. The number of such dipoles determines the magnitude of the field and increases linearly with increasing field strength. The vector representing the intensity of polarization exhibits a maximum value for those values of applied electric field for which all the molecules are oriented in the direction of the electric field. Under weak applied fields, complete orientation is prevented by thermal agitation of the molecules. In view of this, the degree of polarization is strongly dependent on temperature.

The higher the temperature, the greater is the thermal agitation displacing the particles subject to the external electric field. Therefore, the polarizability of such particles must decrease with increasing temperature. The relationship between polarizability, α_d, of the molecules and the temperature is given by the following equation:

$$\alpha_d = \frac{\mu^2}{3kT} \, ,$$

where μ is the electric moment of the polar molecules, k is Boltzmann's constant, and T is the absolute temperature.

The relation between polarizability and dielectric permeability for polar gases and much-diluted polar liquids resembles the Clausius–Mosotti relation and is called the Debye formula:

$$\frac{\varepsilon - 1}{\varepsilon + 1} \cdot \frac{M}{\rho} = \frac{4}{3} \, \pi N \left(\alpha_e + \frac{\mu^2}{3kT} \right) . \tag{II.1a}$$

Here, M is the molecular weight of the material, ρ is the density, N is Avogadro's number, and α_e is the polarizability caused by electron displacement.

For concentrated polar liquids, the Onsager relation is applicable

$$\frac{(\varepsilon - \nu^2)(2\varepsilon + \nu^2)}{\varepsilon (\nu^2 + 2)^2} = 4\pi n \frac{\mu^2}{9kT},$$

where ν is the index of refraction for the pure liquid consisting of polar molecules. Knowing the dipole moment for the molecules and the index of refraction of the liquid, it is possible to compute the dielectric permeability.

Dipole-relaxation polarization is observed in polar gases and liquids, as well as in crystals containing water of crystallization or hydroxyl radicals in the crystal structure.

Ionic-relaxation polarization sometimes is induced by an external electric field if there are weakly bonded ions which have been displaced from their normal lattice positions by thermal motion. If such ions are present, they exhibit a dipole moment (because polarized). Polarization in such dielectrics is significantly temperature-dependent. With increasing temperature, the thermal motion of the ions increases and their polarizability falls.

The polarizability of the particles in ionic-relaxation polarization can be expressed as follows:

$$\alpha_i = \frac{q^2 \delta^2}{12kT},$$

where q is the ionic charge and δ is the mean free path. Ionic-relaxation polarization is observed in crystalline dielectrics when they contain foreign ions or lattice defects.

In materials with electron or "hole" conductivity, electronic-relaxation polarization may be observed. The source is a defect "hole" developed by thermal energy. This type of polarization, as is well known, causes a very high value of dielectric permeability. The dielectric permeability plotted as a function of temperature exhibits a maximum at low temperatures.

A form of relaxation polarization termed structural polarization occurs in inhomogeneous materials. In a material containing inclusions of a conductor or semiconductor, substantial accumulations of charge may be found at the interfaces, leading to anomalously large values of dielectric permeability at low frequencies. At high frequencies, the amount of interfacial polarization is less, and the dielectric permeability drops.

All the various types of relaxation polarization require fairly long excitation and relaxation times. The excitation time required for this type of polarization is much greater than the excitation time required for electron displacement. As a result, materials exhibiting relaxation polarization also exhibit dispersion in dielectric permeability in the range of frequencies of geophysical interest.

We may define a relaxation time, τ, for the various types of relaxation polarization which have been described. It reflects the rapidity with which ions or dipoles become oriented when an external field is applied. The relaxation time specifically is the time required for the electric moment of a unit volume to decrease by the ratio $1/e$ when the external field is removed.

The relaxation time is a function of the activation energy, the natural frequency of oscillation of the polarizing particles, and of temperature. It is given by the equation.

$$\tau = \frac{1}{2U} e^{E_0/kT},$$

where U is the frequency of ionic vibration in its rest position, and E_0 is the activation energy.

Thus, the relaxation time decreases with increasing natural frequency and increasing temperature, but increases with increasing activation energy for the polarizing particles.

In a real dielectric material, polarization may result from several different polarizing processes. In the most general case, the electric moment caused by all forms of polarization is the product of the number of polarizing particles, the intensity of the applied electric field, and the algebraic sum of all the individual polarizabilities.

The presence of one or another of these types of polarization depends on the physicochemical properties of the material and frequency range under consideration, inasmuch as the time required for polarization varies widely among the different processes. Howell and Licastro [14] have given a schematic though illuminating description of the variation of dielectric permeability with frequency for a material which exhibits structural, dipole, atomic, and electronic polarizations (Fig. 1). With increasing frequency, the dielectric permeability decreases. This change of character

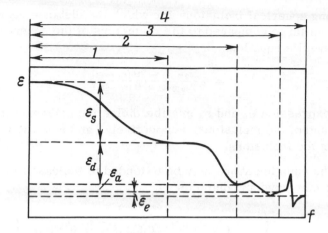

Fig. 1. Idealized relationship between dielectric constant and frequency for various forms of polarization (from Howell and Licastro): (1) interfacial (ε_s), (2) dipole (ε_d), (3) atomic (ε_a), (4) electronic (ε_e).

is explained by the decrease in polarization when the external field resonates the polarizing particles.

In rocks, which are multiphase systems made up of minerals with different compositions, most of these types of polarization take place, with the result that dielectric permeability is a complicated function of frequency, temperature, pressure, and other factors.

The dielectric permeability of a rock may be computed from data on the composition of the material and the dielectric permeabilities for the various individual components comprising the material. In order to do this, a rock must be considered as either a matrix or a mixture, and different equations must be used in the two cases. *

There are a variety of equations available for computing the dielectric permeability of a matrix [15-17]. In the case of a matrix

---

*In a matrix, one of the components serves as a continuous host material, while the other component occurs as isolated particles within this matrix. An example is a rock in which sulfide minerals occur as disseminated blebs, or a rock which is filled with air or water.

containing spherical inclusions for which the distance between spheres is large compared to the diameters of the spheres, we may use Maxwell's equation

$$\varepsilon = \varepsilon_0 \frac{2\varepsilon_0 + \varepsilon_1 - 2\Theta(\varepsilon_0 - \varepsilon_1)}{2\varepsilon_0 + \varepsilon_1 + \Theta(\varepsilon_0 - \varepsilon_1)}.$$

In this expression ε_0 and ε_1 are the dielectric permeabilities of the matrix and of the inclusions, respectively, and Θ is the volume ratio for the inclusions.

The equation above may be extended to the case in which the spheres may have any diameter

$$\varepsilon = \varepsilon_0 \left(1 - \frac{3\Theta}{\Theta + \dfrac{2\varepsilon_0 + \varepsilon_1}{\varepsilon_0 - \varepsilon_1} - 0.523 \dfrac{\varepsilon_0 - \varepsilon_1}{4/3\varepsilon_0 + \varepsilon_1} \Theta^{10/3} + \cdots}\right).$$

This expression is more accurate than the expression given by Maxwell, inasmuch as the assumption that the spherical inclusions are relatively small need not be made.

In computing the dielectric permeability for a system which is an unordered mixture of two components, the best agreement with experimental data is given by Lichteniker's formula

$$\log \varepsilon = \Theta_1 \log \varepsilon_1 + \Theta_2 \log \varepsilon_2.$$

If there is a large difference in the dielectric permeabilities, Odelevskii's formula provides better results

$$\varepsilon = B + \sqrt{B^2 + \frac{\varepsilon_1 \varepsilon_2}{2}},$$

where

$$B = \frac{(3\Theta_1 - 1)\varepsilon_1 + (3\Theta_2 - 1)\varepsilon_2}{4};$$

ε, ε_1, and ε_2 are the dielectric permeabilities for the mixture and for the components, respectively, and Θ_1 and Θ_2 are the volume fractions for the two components, with $\Theta_1 + \Theta_2 = 1$.

Tarkhov [18] has used the Lorentz–Lorenz formula for computing the average value for ε for a number of water-bearing rocks, with good agreement between computed and measured values

$$\frac{\varepsilon - 1}{\varepsilon + 2} = \frac{\varepsilon_1 - 1}{\varepsilon_2 + 2}\Theta_1 + \frac{\varepsilon_2 - 1}{\varepsilon_2 + 2}\Theta_2 + \cdots$$

All of these expressions apply for mechanical mixtures of materials, and not for cases in which interactions occur between the materials.

It should be noted that water present in the pore structure of rock acts as a strong electrolyte and reacts with the solid rock framework.   This water as well as anisotropy may lead to significant differences between calculated and measured values of dielectric permeability.   Therefore, it is highly desirable to have measured values of ε and the possible range of values over the frequency spectrum which is of concern.

Dielectric Permeability of Minerals

Data concerning the dielectric permeability of minerals are becoming more valuable with the increased interest in the behavior of AC fields in rocks, and such data are in themselves interesting. The difference in dielectric permeabilities for various minerals may be used as a basis for separating such minerals.   Minerals such as mica, quartz, asbestos, and talc are widely used as insulating materials.   Knowledge of the values for dielectric properties is necessary in the application of electrical prospecting methods using alternating currents, such as are being used in exploration geophysics.   It must be admitted that the study of the dielectric properties of minerals as well as those of rocks has not been given enough attention.   Most work to date has been done with mica and quartz, which are used industrially.   For almost all of the other minerals, we have only one or two measurements for each, which were made with different techniques at different frequencies and with doubtful accuracy.   As a result, we find widely different values for the same mineral from different sources.   Moreover, the dielectric permeability of the same mineral specimen may vary as a result of chemical variations, foreign inclusions, or moisture content.

Values for the dielectric permeability of minerals vary over about two orders of magnitude.   The smallest value observed for a mineral is approximately 3, while the largest value is observed for rutile, being about 173.   The question as to the reason for differences in dielectric permeability among the various minerals may logically be raised.   We will give some general ideas concerning this question in the following paragraphs.

In considering basic data for some pure minerals and single crystals, Povarennykh [19,20] has deduced that the dielectric permeability for ionic covalent bonded minerals increases with increasing atomic weight, the degree of covalency, the coordination number, and the number of free electrons, but decreases with increasing interatomic spacing. Above all, it is apparent that the internal structure and other structure-related factors are primary in determining the size of the dielectric permeability. However, at present there are not enough data to test fully Povarennykh's hypotheses. It should also be noted that some of the reported values for dielectric permeability have not been referred to the crystallographic directions, so that variations in the tensor values which reflect anisotropy are not apparent. The difference between values for dielectric permeability as a function of direction may be quite significant for some minerals [21-23].

Values for the dielectric permeability, chemical formulas, the square of the index of refraction, hardness, and density of various minerals, are listed in Table 2. In addition, the frequency range over which ε was measured is given for some of the minerals. Based on chemical compositions, the minerals have been grouped in the table into the following: elements, sulfides, halides, oxides, hydroxides, acid salts, and silicates.

The square of the index of refraction, the hardness, and the density are given in the right-hand columns in Table 2. Using Maxwell's relation that $\varepsilon = \nu_2$ for electron polarization, and comparing values ε and ν, it is possible to estimate how much polarization takes place in addition to electron polarization.

The hardness of a mineral, expressed on the Mohs scale, is a highly important property, inasmuch as it reflects the energy of the crystal lattice. With increasing binding energy, the hardness increases. However, for minerals such as halite, fluorite, cesium chloride, sphalerite, and wurtzite, there is an inverse relation between molecular polarizability and binding energy. This suggests that there is an inverse relationship between the dielectric permeability and the hardness of minerals: the harder the mineral, the lower should be the dielectric permeability.

Obviously, the dielectric permeability of a material is a function not only of the polarizability of the individual particles, but also of the number per unit volume. This parameter is closely related to the density, chemical composition, and structure of a material.

Moreover, the electrical conductivities of the various minerals should also be listed in Table 2, inasmuch as it is known that electron conduction results in high values of dielectric permeability. On the other hand, a comparatively weak contribution to dielectric permeability is observed in materials with ionic conductivity. However, very few data on electrical conductivity were reported along with the values for dielectric permeability; therefore, they could not be listed.

In comparing the values for dielectric permeability with the other physical properties within each group, we note the following. For minerals consisting of single elements (carbon, sulfur, and selenium), the observed values for dielectric permeability are very nearly equal to the square of the index of refraction. The small differences can probably be attributed to experimental errors in measuring ε. The dielectric permeability of carbon, sulfur, and selenium is apparently contributed solely by electron polarization. Graphite appears not to fall in this category. The high value of dielectric permeability for graphite is probably caused by its semiconductor characteristics. The dielectric permeability of carbon, sulfur, and selenium increases with increasing density.

The metal sulfide compounds are listed in the second group. This group can be divided into two subgroups on the basis of the value of dielectric permeability – those which have values in the range of 20 or higher, and those with dielectric permeabilities of 6 to 8. For the latter group, the dielectric permeability is not greatly different from the square of the index of refraction; therefore, electron polarization must be the important mechanism. The silver compounds, which are characterized by dielectric permeabilities greater than 81, constitute a smaller subgroup. The high values for such compounds result from the fact that the radius of the silver ion is very large and has a high polarizability ($\alpha = 5.9 \cdot 10^{-24}$ cm^3). Silver ions form homologous compounds more readily than the acid compounds which most minerals are formed as. Moreover, frequently the metals which combine with silver forming compounds in the first subgroup appear as strongly polarizable ions imbedded in an electron cloud with 18 electrons. For example, lead has a polarizability $\alpha = 4.32 \cdot 10^{-24}$ cm^3. Somewhat

Table 2. Dielectric Permeability, ε, and Several Other Physical Properties of Minerals

Mineral	Chemical formula	ε	Frequency, cps	Index of refraction squared	Hardness Mohs scale	Density, g/cm³	Reference
Native elements							
Carbon	C	5.7		5.76	10	3.51	[19]
Sulfur	S	4.1		4.2	2	2.07	[19]
Selenium	Se	10.7		11.2	2	4.5	[19]
Graphite	C	>81		4.0	2	2.15—2.25	[19]
Sulfides							
Chalcocite	Cu_2S	>81			2.5—3	5.5—5.8	[19]
Argentite	Ag_2S	>81			2—2.5	7.2—7.4	[19]
Galena	PbS	>81			2.5	7.57	[19]
Pyrrhotite	from Fe_6S_7 to $Fe_{11}S_{12}$	>81			3.5—4	4.77	[19]
Molybdenite	MoS_2	>81		4—18.5		4.92	[19]
Pyrite	FeS_2	>33.7<81			6—6.5	4.9—5.02	[19]
Cobaltite	CoAsS	>33.7<81			5—6	6.0—6.5	[19]
Arsenopyrite	FeAsS	>81			5.5—6	5.9—6.2	[19]
Cinnabar	HgS	6.2	50	7.3—10.7	2.5	8.1	[21]
Arseniopyrite	As_2S_3	7.2		5.7—9.0	2	3.4—3.5	[19]
Realgar	AsS	7.6	50	6.2—7.2	1.5—2	3.58	[21]
Spalerite	ZnS	7.8		5.7	3—4	3.5—4.2	[22]
Wurtzite	ZnS	8.2		5.75	3.5—4	3.98—4.08	[19]
Antimonite	Sb_2S_3	11.2		9.6—18.5	2—2.5	4.5—4.6	[19]
Halides							
Halite	NaCl	5.6—6.4		2.4	2.5	2.1—2.2	[24]
Sylvite	KCl	4.68—4.8		2.2	2.0	1.97—1.99	[24]
Cerargyrite	AgCl	12.3		4.3	1.5—2.0	5.55	[12]

Mineral	Formula						Ref.
Fluorite	CaF_2	3.0–3.2	4.0	2.05		6.2–8.5	[21]
Cryolite	Na_3AlF_6	2.25–3	2.3	1.7		8.1	[19]
Oxides							
Chalcedony	SiO_2	2.55–2.63	6	2.39	$5 \cdot 10^1 - 5 \cdot 10^1$	5.6–7.5	[24]
Cuprite	Cu_2O	5.85–6.15	3.5–4	8.3	50	5.65–6.35	[21]
Tenorite	CuO	5.8–6.4	3.5	6.8		13.1	[19]
Zincite	ZnO	5.4–5.7	4	4.0		11	[19]
Spinel	$MgAl_2O_3$	3.55–5.04	7–8	2.9		6.8	[19]
Hematite	Fe_2O_3	5.3	5.5–6.5	9.0		25	[22]
Magnetite	Fe_3O_4	4.9–5.2	5.5–6.5	5.6		$>33.7<81$	[19]
Ilmenite	$FeTiO_3$	4.44–5.0	5–6.0	4.0		$>33.7<81$	[19]
Franklinite	$(Mn, Zn)\,Fe_2O_4$	5.82	5.5	5.6		6.4	[19]
Chromite	$FeCr_2O_4$	5.09	5.5–8	4.5		11.0	[19]
Braunite	Mn_2O_3	4.8	6–6.5			>81	[19]
Pyrolusite	MnO_2	4.7–5.0	2–6	6.25		10^5	[25]
Lead monoxide	PbO	9.13	2	5.3		26	[19]
Platnerite	PbO_2		5.5	6.25		26	[19]
Brookite	TiO_2	3.67–4.01	5.5–6			78	[9]
Anatase	TiO_2	3.9	5.5	1.96		48.0	[19]
Rutile	TiO_2	4.2	6–6.5	8.4	50	89–173	[21]
Perovskite	$CaTiO_3$	4.0	5.5	5.6	50	170	[21]
Samarskite	$(\dots N_6Ta)_2O_6$	5.6–5.8	5–6	4.0		7.7	[21]
Ceramic "Vinkit"	$(N_6, Ta, Ti)_2O_6$	4.7–5.4	5–6		50	6	[21]
Ceramic "Éshinit"	$(Ce, Fe, Ca) - - (Nb, Ti, Th)_2O_6$	5.1	5.5	4.41	50	5.8	[21]
Ceramic "Évksenit"	$(Y, Ca, Ce) - - (Nb, Ta, Ti)_2O_6$	4.8–5.9	5–6	4–5.29	50	5.2	[21]
Cassiterite	SnO_2	6.8–7.0	6–7			24.0	[19]

Table 2. (continued)

Mineral	Chemical formula	ε	Frequency, cps	Index of refraction squared	Hardness, Mohs scale	Density, g/cm³	Reference
Hydroxides							
Cliachite	$Al(OH)_3$	8.4		2.5	3	2.3–2.42	[-9]
Diaspore	$HAlO_2$	6.2		2.9	6.5–7	3.3–3.5	[19]
Manganite	$MnO_2Mn(OH)_2$	>8.1		5.0	3–4	4.2–4.4	[19]
Goethite	$HFeO_2$	11.7		5.3	5–5.5	4.0–4.4	[19]
Limonite	$HFeO_2 \cdot aq$	3.2	50	4.0	4.0	3.5–4.0	[21]
Acid salts,							
Carbonates							
Calcite	$CaCO_3$	7.5–8.7		2.2–2.8	3	2.6–2.8	[24]
Aragonite	$CaCO_3$	7.4		2.38–2.8	3.5–4	2.9–3.0	[19]
Magnesite	$MgCO_3$	10.6	50	2.2–2.9	4.0–4.5	2.9–3.1	[21]
Siderite	$FeCO_3$	5.2–7.4	50	2.6–3.5	3.5–4.5	3.9	[21]
Cerussite	$PbCO_3$	22.7		4.27	3–3.5	6.4–6.6	[19]
Strontianite	$SrCO_3$	7.0		2.8	3.5–4	3.6–3.8	[19]
Witherite	$BaCO_3$	5.7		2.8	3–3.5	4.2–4.3	[19]
Rhodochrosite	$MnCO_3$	6.8		3.5	3.5–4.5	3.6–3.7	[19]
Dolomite	$CaMg[CO_3]_2$	6.3–8.2		2.2–2.8	3.8–4	1.8–2.8	[21]
Ankerite	$Ca(Mg,Fe)[CO_3]_2$	7.0		2.2–2.9	3.5–4.5	2.96–4.1	[21]
Azurite	$Cu_3(OH)_2(CO_3)_2$	6.0	50	3.0	3.5–4	3.8	[21]
Malachite	$Cu_2(OH)_2[CO_3]$	4.4	50	3.3	3.5–4.0	3.9–4.5	[21]
Sulfates							
Barite	$BaSO_4$	6.2–7.9		2.7	3	4.5	[21]
Celestite	$SrSO_4$	7.0		2.7	3.5	4.0	[19]
Anglesite	$PbSO_4$	14.0		3.5	2.5–3	6.3	[19]
Anhydrite	$CaSO_4$	6.5		2.46	3	2.8–3.9	[19]

Mineral	Formula						Ref.
Gypsum	$CaSO_4 \cdot 2H_2O$	7.9–6.3		2.3	1.5–2.0	2.31–2.33	[21]
Mirabilite	$Na_2SO_4 \cdot 10H_2O$	8.3		1.96	1.5–2·	1.46	[21]
Chalcanthite	$CuSO_4 \cdot 5H_2O$	7.8		2.3	2.5	2.2	[19]
Astrakhanite	$Na_2Mg(SO_4)_2 \cdot 4H_2O$	6.5		2.2	3	2.23	[21]
Epsomite	$MgSO_4 \cdot 7H_2O$	6.7		2.1	2–2.5	1.77	[21]
Tungstates and molybdenites							
Wolframite	$(Mn, Fe)WO_4$	14		5.3	4.5–5.5	7.18	[21]
Scheelite	$CaWO_4$	3.5		3.6	4.5–5	6.12	[21]
Wulfenite	$PbMoO_4$	26.8		5.6	3	6.9	[19]
Phosphates							
Apatite	$3Ca_3P_2O_8Ca(F, Cl)_2$	5.8		2.5	5	3.15–3.27	[21]
Pyromoerphite	$Pb_5(PO_4)_3Cl$	26		4.2	4	6.7–7.1	[19]
Monazite	$(Ce, La)PO_4$	3–6.6		3.1	5	4.95–5.3	[21]
Vivianite	$Fe_3(PO_4)_2 \cdot 8H_2O$	8.2	50	2.56	2	2.95	[21]
nitrates and chromates							
Niter	KNO_3	5		2.9	2	1.99	[19]
Crocoite	$PbCrO_4$	9.6		5.6	2.5–3	6.0	[19]
Silicates							
Zircon	$ZrSiO_4$	3.6–5.2		3.6	7.5	4.02–4.86	[21]
Olivene	$(Mg, Fe)_2SiO_4$	6.8		2.56	6.5–7	3.18–3.57	[19]
Topaz	$Al_2(F, OH)_2[SiO_4]$	7.4		2.56	8	3.4–3.65	[21]
Vesuvianite (idiocrase)	$Ca_3Al_2[SiO_4]_2[OH]_4$	7.2		2.9	6.5	3.34–3.44	[21]
Kyanite	Al_2OSiO_4	5.7	50	2.9	7.5	3.56–3.68	[21]
Actinolite	$Ca_2(Mg, Fe^{\cdot\cdot})_5-[Si_4O_{11}]_2[OH]_2$	6.6				3.1–3.3	[21]
Serpentine	$Mg_6[Si_4O_{10}][OH]_8$	10.0	50		2.5–3	2.2–5.7	[21]
Chloritoid	$Fe_2^{\cdot\cdot}Al_2[Al_2Si_2O_{10}] \cdot [OH]_4$	6.9	50	2.89	6.5	3.26–3.57	[21]

Table 2. (continued)

Mineral	Chemical formula	ε	Frequency, cps	Index of refraction squared	Hardness, Mohs scale	Density, g/cm³	Reference
Anthophyllite	$(Mg, Fe)_7(OH)_2Si_8O_{22}$	10.5	50	2.72	5.6-6	2.9-3.4	[21]
Steatite	$Mg_3(OH)_2Si_4O_{10}$	7.9	50	2.25-2.56	1	2.82	[21]
Heulandite	$(Ca, Na_2)[AlSi_3O_8]_2 \cdot 5H_2O$	7.5	50	2.25	3.5-4	2.2	[21]
Dioptase	$Cu_6Si_6O_{18} \cdot 6H_2O$	7.6		2.9	5	3.18-3.35	[21]
Diopside	$CaMg[Si_2O_6]$	10.0		2.9	5-6	3.27-3.38	[21]
Hedenbergite (diopside)	$CaFe^{\cdot\cdot}[Si_2O_6]$	9.0		2.9	5-6	3.5-3.6	[19]
Aegerite	$NaFe^{\cdot\cdot}[Si_2O_6]$	7.2		2.9	5-6	3.43-3.60	[21]
Spodumene	$LiAl[Si_2O_6]$	8.4		2.7	6-7	3.13-3.20	[19]
Actinolite	$Ca_2(Mg, Fe^{\cdot\cdot})_5[Si_4O_{11}]_2 - [OH]_2$	6.6			5-6	3.1-3.3	[21]
Hornblende	$Ca_2Na(Mg, Fe^{\cdot\cdot})_4 - (Al, Fe^{\cdot\cdot\cdot})[Si_3AlO_{11}]_2$	4.9-5.8			5-6	3.1-3.37	[21]
Wollastonite	$Ca_3[Si_3O_9]$	6.2		2.5	4.5-5	2.78-2.91	[19]
Rhodonite	$(Ca, Mn)SiO_3$	4.6		2.9	5.5-6.5	3.40-3.75	[21]
Prehnite	$Ca_2Al(OH)_2AlSi_3O_{10}$	6.5		2.7	6-6.5		[19]
Desminite	$(Ca, Na_2)Al_2Si_6O_{16} \cdot 6H_2O$	6.6		2.25	3.5-4	2.09-2.20	[21]
Sphene	$CaTiOSiO_4$	4.0-6.6		4.0	5-5.5	3.29-3.56	[21]
Almandite	$Fe_3Al_2[SiO_4]_3$	4.3		2.7	-	3.5-4.2	[21]
Andradite	$Ca_3Fe_2[SiO_4]_3$	8.2		3.2	5.7-6.5		[21]
Grossularite	$Ca_3Al_2Si_3O_{12}$	7.6		3.2	3-3.5		[19]
Spessartite	$MnAl_2Si_3O_{12}$	7.6		3.2	7.5		[21]
Epidote	$Ca_2(Al, Fe)_3[Si_2O_7][SiO_4]O[OH]$	6.2		3.0	6-7	3.35-3.45	[21]
Apatite	$Ca, Ce, La)_2(Al, Fe)_2 - [OH]O[Si_2O_7][SiO_4]$	4.2		3.1	6	4.1-2.7	[21]

Mineral	Formula						
Beryl	$Be_3Al_2[Si_6O_{18}]$	3.9−7.7		2.58	7.5−8	2.63−2.91	[2]
Cordierite	$Al_3(Mg, Fe)_2[AlSi_5O_{18}]$	7.0		**2.36**	7−7.5	2.60−2.66	[15]
Chrysocolla	$CuSiO_3 \cdot 2H_2O$	13.1		2.66	2	2.0−2.3	[21]
Tremolite	$Ca_2Mg_5[Si_4O_{11}]_2(OH)_2$	7.6		2.56	5−6	2.9−3.0	[2]
Natrolite	$Na_2[Si_3O_{10}Al_2] \cdot 2H_2O$	5.5		2.18		2.2−2.5	[21]
Augite	$Ca(Mg, Fe, Al) -$ $[Si, Al_2O_6]$	6.8		2.8	5−6	3.2−3.6	**[19]**
Talc	$Mg_3[Si_4O_{10}][OH]_2$	5.8		2.7	1	2.7−2.8	[21]
Vermiculite	$(Mg, Fe^{\cdot\cdot}, Fe^{\cdots})_3 -$ $[Si, Al_4O_{10}][OH]_2 \cdot 4H_2O$	9.5−13.5	50	2.4	1.5	2.4−**2.7**	[21]
Phlogopite	$KMg_3[Si_3AlO_{10}][Fe, OH]_2$	7.0	50	2.5	2−3	2.70−2.85	[21]
Biotite	$K(MgFe)_3[AlSi_3O_{10}] -$ $(OH, F)_2$	10.3		2.56	2−3	2.70−3.16	[21]
Muscovite	$KAl_2[OH]_2[AlSi_3O_{10}]$	6.2−8.0		2.56	2−3	2.76−3.10	[24]
Fuchsite	the same, with Cr substituted for Al	5.0	50	2.56	2−3	2.76−3.10	[21]
Lepidolite	$KLi_{1.5}Al_{1.5}[AlSi_3O_{10}] -$ $[F,OH]_2$		50	2.4	2.5−4	2.8−2.9	[21]
Kaolinite	$Al_4[OH]_8[Si_4O_{10}]$	9.1	50	2.4	1−2.5	2.60−2.63	[21]
Albite	$Na[AlSi_3O_8]$	6.0		2.5	6−6.5	2.62	[19]
Labradorite	$Na_2Ca_3[AlSi_3O_8]_8$	5.8		2.5	6−6.5	2.67−2.72	[21]
Orthoclase (feldspar)	$K[AlSi_3O_8]$	4.5−6.2	10^2-10^7	2.3	6−6.5	2.5−2.62	[24]
Microcline (feldspar)	$K[AlSi_3O_8]$	5.6	50	2.2	6−6.5	2.54−2.57	[21]
Leucite	$K[AlSi_2O_6]$	6.8		2.2	5.5−6	2.45−2.50	[5]
Nepheline	$Na[AlSiO_4]$	6.2	50	2.3	5.5−6	2.55−2.65	[2]
Sodalite	$Na_8[AlSiO_4]_6Cl_2$	6.8		2.2	5.5−6	2.35	[5]
Analcime	$NaAlSi_2O_6 \cdot H_2O$	5.8		2.2	5−5.5	2.2−2.3	[2]
Anorthite	$Ca[Al_2Si_2O_8]$	6.9	50	2.5	6−6.5	2.76	[19]

smaller, though still large, polarizabilities are observed for copper and silver ($\alpha_{Cu} = 1.8 \cdot 10^{-24}$ cm^3, $\alpha_{Ag} = 1.85 \cdot 10^{-24}$ cm^3). Combining the sulfur anion with a cation having high polarizability leads to a material with a very high dielectric permeability.

A close relationship between the high values of dielectric permeability and hardness on one hand and density on the other hand is not apparent. The division of minerals according to the value of dielectric permeability is also a subdivision in terms of conductivity.

Most of the halide compounds are characterized by ionic bonding. The dielectric permeability is considerably larger than the square of the index of refraction for these minerals. Therefore, there must be a considerable amount of ionic polarizability in these minerals in addition to the electronic polarizability. For some of the minerals in this group, the dielectric permeability increases with increasing density.

Minerals which are oxide compounds are the most widely distributed minerals in the earth's crust. Simple metal oxide compounds are included in the groups of salts and hydroxides. Their physical properties are determined fundamentally by the size and valence of the cations, as well as the chemical relationship between the ions, inasmuch as the size of the anions O^{-2} and $(OH)^{-1}$ is practically the same. In the crystal structure of these compounds, the cations always occur within an enveloping swarm of oxygen anions, and the coordination number for the crystal structure is an essential characteristic of these minerals. Povarennykh [19] gives a series of examples indicating that the dielectric permeability increases with increasing atomic coordination number. This relationship requires further investigation.

The oxide and hydroxide minerals, as was the case for sulfides, are characterized by high values for dielectric permeability. Oxides with $\varepsilon < 6$ are excluded; most values for dielectric permeability are considerably higher. Electron polarization appears to be the sole polarization mechanism only for the mineral cuprite and the ceramic evksenit. For the other minerals, other forms of polarization appear to contribute to the value of ε. It is obvious that the minerals with high dielectric permeability in this group are mainly those with oxygen ions. Oxygen has a high polarizability ($\alpha = 2.76 \cdot 10^{-24}$ cm^3) although it is less polarizable than the sulfate radical, which is offset by the large value of

$\alpha/r = 1.2$, while it is only 1.12 for sulfur. The ratio of polarizability to ion spacing, α/r, plays an important role, inasmuch as
the dielectric permability is a function not only of the polarizability of the elementary particles, but also of particles per unit volume. Thus, the less the radius r for particles with a high polari-
ability, the greater must be the dielectric permeability.

The oxides of ambivalent metals are characterized by particularly high values of dielectric permeability, as, for example,
platnerite, hematite, magnetite, and ilmenite. The largest values
of ε are obtained for compounds of manganese (braunite, pyrolusite) and titanium (rutile, perovskite, anatase, and brookite). It
has been established that the crystal structure is the cause of the
high dielectric permeability. The combination of ions in the rutile
or perovskite lattice generates an internal electric field which
reinforces the external field. The enhanced internal field increases the polarization and explains the high value for dielectric
permeability.

The oxide minerals are characterized by moderately high
values of the index of refraction ν and by high values for hardness
(about 5.5)and density in comparison with the broad group of acid
salts.

The particular characteristic of an acid salt is that it con-
tains an ion complex, such as $(CO_3)^{-2}$, $(SO_4)^{-2}$, $(PO_4)^{-3}$, $(NO_3)^{-1}$,
and others, in the crystal lattice as a structural entity. The cation
in such a complex has a small ionic radius and a high charge and
is bonded covalently to the oxygen atoms. The structure of an acid
salt is typically that of an ionic compound. Therefore, the crystal
structure fundamentally must be determined by the relative sizes
of the cations and anions, taken as structural units.

The acid salts listed in Table 2 are subdivided into several
subgroups on the basis of the chemical composition of the anion
complex. For the subgroups of carbonates, sulfates, tungstates,
and phosphates, minerals have values for dielectric permeability
typically in the range 6 to 8. The only exceptions are salts in which the
cation contains lead. These are the minerals cerussite, anglesite,
croicite, wulfenite, and pyromorphite, as well as the mineral
wulframite which contains the element manganese as a cation. The
dielectric permeability of these minerals is quite high, in the
range 10 to 26. The fact that the values for dielectric permeability
for minerals in this group differ significantly from the square of

the corresponding index of refraction indicates that mechanisms
of polarization other than electron displacement must be important.
No relation between dielectric permeability and hardness is found,
but there is a relation with density. All minerals with high values
for ε have significantly higher densities.

The last and most complex in chemistry of the groups of
minerals is the silicate group. Their structures are determined

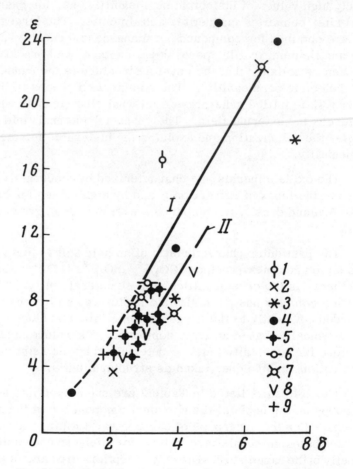

Fig. 2. Relationship between dielectric constant of minerals
and their density compiled by V. N. Kobranova, using data
from several authors [24]: I — Minerals with ionic crystal
structure; II — Silicates. (1) Diamond, (2) sulfur, (3) sulfides, (4) oxides, (5) silicates, (6) phosphates, (7) carbonates,
(8) sulfates, (9) chlorides.

by a basic silicate or aluminosilicate structure which exhibits a
covalent crystal bond. For most such minerals, the value for
dielectric permeability falls in the range 6 to 8. The highest values,
8 to 10, are observed for a small number of minerals which contain ions such as Fe or Mg. A few minerals have values for ε of
the order of 4.

The square of the index of refraction for these minerals falls
in the range 2.2 to 4.0, but a large number of the minerals have
$v^2 \cong 2.5$. This is only half the value of the dielectric permeability.
Therefore, it is reasonable to assume that ion displacement polarization and possibly other mechanisms are important. The data
shown in Fig. 2 indicate that the value for dielectric permeability
increases with increasing density, but not as markedly as for some
of the oxides.

Considering the generalizations which were made in the preceding paragraphs, it appears that pure electronic polarization
takes place in a few minerals (sulfur, selenium, and carbon). The
dielectric permeability of most minerals with ionic covalent bonding is less than 12. Most of the high values of dielectric permeability are found in the oxide and sulfide groups. Many of the
minerals in these groups contain heavy metal cations such as Pb^{+2},
Mn^{+3}, or Re^{+3} (magnetite, hematite, braunite, and so on) or have
the structure of rutile or perovskite (pyrolusite), which leads to a
high value of dielectric permeability. In some of these minerals,
in addition to electron displacement polarization, an important
role is assumed by electronic relaxation polarization which occurs
as the result of thermal activation of defect electrons or holes.

Naturally formed minerals commonly contain impurities of a
different ion which are bonded with less energy than the proper
ions in the crystal lattice. As a result, ionic relaxation polarization takes place in these minerals, which causes the value for
dielectric permeability to be temperature dependent.

Within each of the mineral classes, we should note the existence of a direct relationship between dielectric permeability and
density.

It is necessary to subdivide the mineral groups more finely
on the basis of chemical composition and structure in order to
observe correlations between dielectric permeability and hardness.

The range of values for dielectric permeability for a single
mineral from different localities may be explained not only by
differences in the measurement technique and impurities in the
samples, but also by the nature of the impurities. Experimental
data indicate that moisture content is highly important. Primarily,
this importance arises from the fact that the dielectric permeability
of water is 81, which is considerably larger than the value for most
minerals. Table 3 lists values for the dielectric permeability of
three minerals measured as water was being driven from the
samples.

The presence of 0.1 to 0.12% water in apatite and almandine
increases the dielectric permeability to 7.8 and to 4.0, respectively, while in the case of dolomite, a moisture content of 0.38%
increases the dielectric permeability by 13% over that in a dry
state. In view of this dependence of dielectric permeability on
moisture content, it is important to determine moisture content
when ε is determined.

The values for dielectric permeability listed in Table 2 correspond to arbitrary directions. As was noted before, the dielectric properties of minerals should be written in tensor form. This
tensor is symmetrical, and if an orthogonal coordinate system
parallel to the crystal axes is chosen, it will consist of three terms.
Crystal which are one-dimensional are characterized by two constants, while two-dimensional crystals are characterized by three
constants.

If we measure the dielectric permeability of a crystal as a
function of direction and then plot it as a function of direction, we

Table 3. Variation of ε for Minerals* Being Desaturated
at 110-120°C [21]

Mineral	ε for Natural water content of mineral	Water content, %	Value for ε after drying for the period	
			30 min	120 min
Apatite	4.87	0.10	4.5	4.5
Almandine	5.8	0.12	5.34	5.56
Dolomite	8.5	0.38	7.47	7.43

*Coarsest minerals which are reported in this table: 42 to 63 μ.

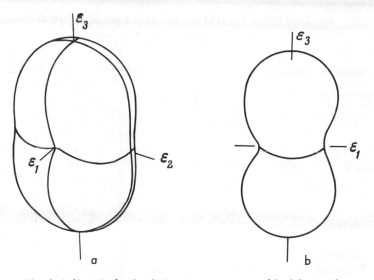

Fig. 3. Indicatrix for the dielectric constant: a — biaxial crystal,
b — uniaxial crystal.

obtain a surface representation of the dielectric permeability which
is called an indicatrix. For one-dimensional crystals, the ε-in-
dicatrix is a geometric figure obtained by rotation about an axis;
for a two-dimensional crystal, the indicatrix has a more complicated form (see Fig. 3).

Table 4 lists values of ε for several minerals measured in
the principal crystallographic directions. The data in the table
indicate the most one-dimensional minerals have a maximum
value for the dielectric constant in a direction normal to the optical axis. A high degree of anisotropy is apparent for the minerals
apatite, augite, gypsum, and barite. The largest range in values
(between ε_1 and ε_2) is nearly 30% in the case of barite. The
dielectric constant for one particular mineral, but with samples
from several sources, may have a variety of values depending on
measurement technique, as well as there being a variety of values
as an inherent property of the mineral. For example, a correlation between ε and density is apparent for corundum. However,
such a correlation is not found in the case of tourmaline [23].

The study of the dielectric properties of minerals is still
far from thorough. There are many minerals, even well known

Table 4. Dielectric Constants of Minerals Measured Along
the Principal Crystallographic Direction

Mineral	ε_1	ε_2	ε_3	Reference
One-dimensional minerals				
Apatite	9.5		7.4	[22]
"	10.2		8.6	[23]
Beryl	7.85		6.05	[22]
"	7.6		6.8	[23]
Dolomite	7.8		6.8	[22]
Calcite	8.49		7.56	[22]
"	8.7		8.2	[23]
Quartz	4.49		4.55	[21]
"	4.69		5.06	[21]
Corundum	9.5—9.8		10.9—11.3	[23]
Rutile	89		173	[12]
Tourmaline	7.13		6.54	[22]
"	6.8—7.9		5.9—6.54	[23]
Cancrinite	9.5		11.2	[26]
Zircon	12.6		12.85	[23]
Quartz, var. citrine	4.55		4.7	[23]
Iceland spar	8.5		8.0	[22]
"	8.49		7.56	[21]
Two-dimensional minerals				
Augite	8.6	6.9	7.1	[18]
Barite	10.0	7.1	7.6	[18]
"	8.1	13.2	8.2	[23]
Gypsum	11.2	12.0	5.4	[18]
Sulfur	3.6	3.9	4.6	[23]
"	.3.52	3.79	4.48	[23]
Topaz	6.65	6.7	6.3	[18]

minerals such as the pyroxenes and plagioclases, for which there
are not data on dielectric constant.

Dielectric Constants of Rocks in Relation to
Mineral Composition

Rocks are complicated materials, both in composition and
structure, so that it is not surprising to find no simple relationship between their physical properties and their chemical or mineral composition. As we have seen, the values for the dielectric
constants of minerals vary over two orders of magnitudes. This

is an adequate variability, so that one might expect some differentiation between rock types on the basis of bulk dielectric constant. In addition to minerals, rocks may contain water and in
some cases, even oil. The amount of liquid in a rock affects the
dielectric constant and, in some instances, plays a dominant role,
particularly at low frequencies. Table 5 lists values for the dielectric constant of rocks with various amounts of moisture.

The common sedimentary rocks are comprised mostly of the
rock-forming minerals calcite, dolomite, and quartz. The dielectric constant of calcite and dolomite is some 1.5 times larger than
the dielectric constant of quartz. As a result, rocks which consist mostly of calcite or dolomite have a dielectric constant which
is higher than that of rocks which consist mostly of quartz, such
as a sandstone. The upper limit for the value of the dielectric constant for a sandstone must be a function basically of the presence
of moisture or minerals with a high value for ε.

The same basic character of behavior is found for the dielectric constant of metamorphic rocks as for sedimentary rocks.
Quartzite has a lower dielectric constant than marble.

Igneous rocks differ widely in composition in comparison
with sedimentary and metamorphic rocks which are relatively consistent in composition. They may consist of two, three, or even
four principal rock-forming minerals. Therefore, considering the
variation in the content of a single mineral may not be adequate
to explain variations in dielectric constant. The primary factor
to consider in the case of igneous rocks is the relative proportions
of acidic and basic minerals.

It is well known that the dielectric constant is greater for
basic and ultrabasic rocks than for acidic rocks. As seen from
the data listed in Table 5, the value for ε increases with increasing content of minerals in the olivene and pyroxene groups. Moreover, for the basic rocks formed at greater depths, the heavy
mineral content is greater, significantly increasing the dielectric
constant. According to data reported by Dostovalov [32, 33], there
is a correlation between dielectric constant and glass content in
andesitic basalts and dacites from Armenia. The curves shown
in Fig. 4 indicate that the dielectric constant increases with decreasing glass content. This correlation indicates that decreasing glass content reflects an increase in the content of minerals

Table 5. Dielectric Constants of Rocks

Rock	Source, mineral composition in %	ε	Frequency, cps	Water content, %	Reference
Sedimentary Rocks					
Anhydrite with gypsum	anhydrite, 92 gypsum, 8	6.3		dry	[18]
"Chuneski" shale, bentonitic		10–45.0	10^2–10^6	10	[27]
The same		4–7.0	10^2–10^6	dry	[27]
"Gurbrin" shale, bentonitic		9.5–10	10^2–10^6	"	[27]
Dolomite		8.0–8.6	10^3–10^7	"	[27]
Limestone	Georgian SSR	7.3		"	[28]
"		8.0–12.0			[22]
Arkosic sandstone	quartz, 23 feldspar, 75 mica, 2	4.9		dry	[18]
Quartz-feldspar sandstone	quartz, 40 feldspar, 45 other, 15	5.1		dry	[18]
Sandstone	Garm	4.66	$5 \cdot 10^5$	dry	[29]
	Zubovskoe	3.96	$5 \cdot 10^5$	dry	[29]
Variegated sandstone		9.0–11.0		dry	[22]
Shaly sandstone		5.53		0.2	[30]
The same		7.17			
Metamorphic Rocks					
Amphibolite		7.9–8.9	10^5–10^7	dry	[31]

Rock	Locality / composition	Dielectric constant	Frequency	Moisture	Ref.
Gneiss		8.0–15.0	$5 \cdot 10^2 - 5 \cdot 10^7$	dry	[22]
Granite gneiss		8.0–9.0		"	[14]
Quartzite	Shokshinskoe	4.36	$5 \cdot 10^5$		[29]
"	Ridderskoe	4.85	$5 \cdot 10^5$		[29]
"		6.6			[22]
"		7.0			[22]
Marble		8.22		dry	[30]
"		8.37		0.002	[30]
"		8.9–9.0	$10^3 - 10^7$	dry	[14]
Talc slate		7.5–34.0	$5 \cdot 10 - 5 \cdot 10^7$	"	[14]
Micaceous slate		9.0–10.0	$5 \cdot 10 - 5 \cdot 10^7$	"	[14]
Roofing slate	Georgian SSR	6.71		"	[30]
"	"	7.74		"	[30]
Phyllite	"	13.0		0.1	[24]
Igneous rocks, acid					
Biotite-granite aplite	quartz, 40 microcline, 32 plagioclase, 20 other, 8	4.8		dry	[18]
Granite	Altai	5.42	$5 \cdot 10^5$	"	[29]
Granite	Leznikovskoe	4.74	$5 \cdot 10^5$	"	[29]
"	Garm	5.06	$5 \cdot 10^5$	"	[29]
"	Valaamskoe	4.5	$5 \cdot 10^5$	"	[29]
		7.0–9.0		combined moisture	[22]
Volcanic tuff		3.8–4.5		dry	[28]
Igneous rocks, intermediate					
Diorite	Kola Peninsula	5.9–6.3	$10^5 - 10^7$	dry	[31]

Table 5 (continued)

Rock	Source, mineral composition, in %	ε	Frequency, cps	Water content, %	Reference
Dicrite		8.5–11.5	10^4–10^7	dry	[14]
Dacite		6.8–8.16	$3 \cdot 10^6$		[32]
Igneous rocks, basic					
Augite porphyry	Kola Peninsula	9.5–12.6	10^5–10^7	"	[31]
Basalt	Berestovetskoe	15.6	$5 \cdot 10^5$	"	[29]
"	Kutai	10.3	$5 \cdot 10^5$	"	[29]
Gabbro		8.8–10.0	10^4–10^7	"	[14]
Gabbro	Southern Urals (3.26% ore content)	12.8		"	[28]
Diabase	Oneshsk	11.6	$5 \cdot 10^5$	"	[29]
"		9.0–13	10^4–10^7	"	[14]
Labradorite	Ukrainian SSR	7.82		"	[30]
		8.24		0.03	[30]
Igneous rocks, ultrabasic					
Peridotite (Plagioclase)	Kola Peninsula	15.7–18.8	10^5–10^7	dry	[31]
Peridotite		12.1	$5 \cdot 10^5$	"	[29]
"	olivene 70, augite, 25, mica, 5	8.6		"	[18]
Olivene pyroxenite	Kola Peninsula	8.4–9.5	10^5–10^7	"	[31]

Igneous rocks, alkaline	Location	ε	Frequency	Ref.
Syenite		6.83	$5 \cdot 10^5$	[29]
"		13.0 – 14.0		[22]
Aegerine-phyllite nepheline syenite		6.93		[28]
Mica nepheline syenite with aegerine		8.47		[28]
Microcrystalline nepheline syenite		9.55		[28]
Aegerine nepheline syenite		9.56		[28]
Fayalite	near Mariupol	8.32		[28]
Luyavrite	Kola Peninsula	9.7 – 11.4	$10^5 - 10^7$	[31]
Urtite	"	7.3 – 8.5	$10^5 - 10^7$	[31]
Felspathic urtite	Kooshva	11.9		[28]

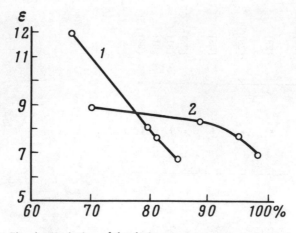

Fig. 4. Variation of the dielectric constant in andesite-basalt (curve 1) and in dacite (curve 2) as a function of the percentage of glass present.

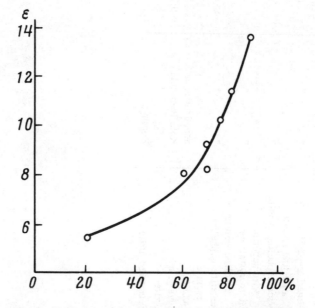

Fig. 5. Variation of the dielectric constant in amphibolite as a function of the percentage of amphibole present [33].

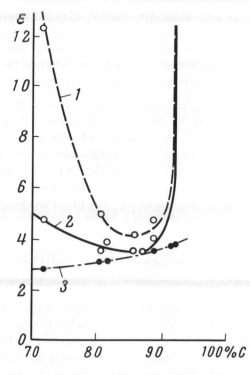

Fig. 6. Variation of the dielectric constant in
wet and dry coals as a function of the carbon
content: (1) air-dried coal, (2) absolutely dry
coal, (3) square of the index of refraction.

with higher dielectric constant. Another example of the correlation between dielectric constant and composition is found in the
case of amphibolite. An increase in amphibole content from 21.4
to 90.7% causes a corresponding incease in dielectric constant
from 5.5 to 13.7 (see Fig. 5).

The dielectric constant of coal, as well as the electrical
conductivity, depends on the carbon ratio of the coal. Groenewage
et al. [34] have studied coals which had undergone various degrees
of metamorphism, and found a correlation between dielectric constant and carbon content which exhibited a minimum at moderate
carbon content. The exact nature of this correlation for carbon
contents up to 80%, as found experimentally (Fig. 6), is strongly
dependent on the moisture content of the coal. Moisture plays a

more important role in low-quality coals, with the dielectric constants of dried coal and coal containing its original moisture being
very nearly the same for a carbon content of 90%. The rapid increase in dielectric constant at high carbon content is associated
with a rapid increase in conductivity as well.

Comparison of measured values for the dielectric constant
of coal with the square of the index of refraction indicates that for
carbon contents below 87%, polarization mechanisms other than
electron displacement must occur. These mechanisms depend on
the presence of polar radicals such as OH and COOH in the coal.
A relationship $\varepsilon = f(C)$ similar to the one discussed here has also
been observed by other investigators [35] for different types of
coal in the frequency range 100 kcps to 30 Mcps.

Effect of Moisture Content on the Dielectric Con-

stant of Rocks

All natural rocks contain some moisture with some degree
of salinity, with the result that the correlation between dielectric
constant and mineral composition may be masked by the correlation between dielectric constant and water content, particularly in
rocks with high water content such as sedimentary rocks.

Even very small amounts of moisture will increase the
dielectric constant markedly. The dielectric constant of water, as
given earlier, is 81, while that of a saline solution is even higher.
The Falkenhagen formula gives the dielectric constant of a solution of a binary electrolyte as follows:

$$\varepsilon = \varepsilon_0 + 3.79 \sqrt{c},$$

where ε is the dielectric constant of the electrolyte, ε_0 is the
dielectric constant of pure water, and c is the concentration of
the electrolyte in moles per liter. For low concentrations, there
is no significant increase in the dielectric constant. Chernyak
[36, 37] has found that the dielectric constant of sand shows a constant value for variation of water salinity over the range 0 to 2
g/liter, and the same relationship was found, $\varepsilon = f(w)$, for sands
saturated with distilled water as for sands saturated with a solution containing low concentrations of NaCl, in the range 1 to 5
g/liter.

Thus, the effect of water salinity on the dielectric constant under these conditions was negligible.

We do not know of any studies which have been made of the dielectric constants of rocks saturated with highly saline solutions. This is because most experiments are concerned with the relation between the dielectric constant and the quantity of water present.

The nature of the relationship between the dielectric constant of a rock and its water content, based on experimenal work [38], depends both on frequency and the degree of water saturation. The frequency dependence of the dielectric constant is relatively minor in comparison with dependence on moisture content. Here, we will consider only the relationship between dielectric constant and the moisture content w for a fixed frequency.

The relationship $\varepsilon = f(w)$ is described by an equation with a form that changes for various degrees of water saturation. For very low water content, measured in tenths or hundredths of a percent, according to Shchodro and Maslova [30], a linear relation $\varepsilon = f(w)$ is observed. By using a very precise measurement technique, that of comparison with a standard, the authors were able to measure very small changes in dielectric constant associated with water content in the range of 0 to 0.1%. In the case of marbles from a few areas, as well as gypsum and granite containing less than a thousandth of a percent water, no decrease in ε was noted on removal of the water content. Variations in ε were observed on removal of water from samples with higher initial water content (in the range up to 0.1%). In these cases, the relationship between ε and w is very nearly linear.

Tarkhov [39] in investigations of the effect of moisture on the dielectric constant divided rocks into three groups: (1) sedimentary rocks with high porosity (from 5 to 15% by volume); (2) igneous rocks with appreciable porosity (from 1 to 5%); and (3) dense rocks (with porosity less than 1%). On the basis of data shown in Fig. 7, he proposed the following empirical formula which is valid for high-porosity sedimentary rocks with water content up to 1%:

$$\varepsilon = \varepsilon_0 w^n, \tag{II.2}$$

where w is the water content of the rock ε_0 is the dielectric constant of the dry rock, and n is an empirical parameter, related to the rock type, with a value falling in the range 0.30 to 0.33. The

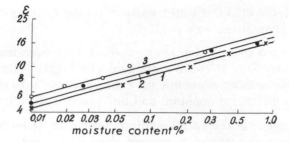
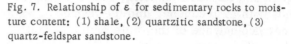

Fig. 7. Relationship of ε for sedimentary rocks to moisture content: (1) shale, (2) quartzitic sandstone, (3) quartz-feldspar sandstone.

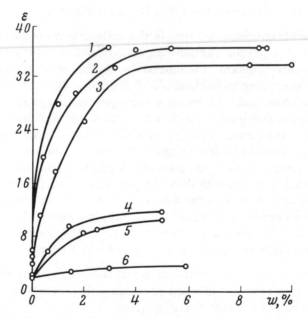

Fig. 8. Relationship of the dielectric constant to moisture content: (1) dolomite, (2) marl, (3) siltstone (data from Parkhomenko and Valeev), (4, 5) siltstone, (6) sandstone (data from Kobranova).

rocks for which data are shown in Fig. 7 are characterized by equal values for the exponent n. However, the same sort of relation is not observed for some other types of rocks. In the case of igneous and metamorphic rocks with porosities of 1 to 5% and water contents up to 0.3%, the value for ε varies with increasing water content, but not in the same manner as was observed for sedimentary rocks for which formula (II.2) was written.

These relationships for the variation of ε with water content are valid only for low water contents (w < 1-2%). Data on the dielectric constant of rocks containing up to 10% water (fig. 8) have been obtained by the author in association with Valeev at a frequency of 10^5cps and by Korennov [38] at a frequency of 10^6 cps. From these data, two typical regions for variation in dielectric constant with moisture content may be recognized. In the first region, for water contents below 3 to 4%, the dielectric constant varies rapidly with increasing water content. The different parts of these curves may be approximated with a series of linear segments or by a single exponential curve.

In the second region, at higher water content, the dielectric constant approaches a limiting value, which varies slowly. A linear relation between ε and w at high water contents (w up to 36%) has been established by Chernyak [36] on the basis of detailed studies of a variety of sands. Rikitaki has found a somewhat similar character for the relation between the dielectric constant of soils and moisture content [40]. According to his measurements, the dielectric constant varies according to a curve which is approximately a parabola from values of about 4 to values of about 30 as the moisture content varies from 0 to 50%.

A significant spread in values for the dielectric constant of similar rocks (curves 3, 4, and 5 on Fig. 8) may be explained by the measurements having been made at different frequencies.

Unequal variations in the dielectric constant with water content between rock types may be explained by the water being present in a bound form in one case and as free water in the other. We must assume that the double layer which is formed at the boundary between solid and liquid phases plays some role. The effect of these factors on the dielectric constant requires further study.

It should be noted that the relation between dielectric constant and water content is more pronounced at audio frequencies than at radio frequencies. Therefore, we may expect a considerable spread in the values for the dielectric constant measured at different frequencies.

Thus, it appears that moisture content is the determining factor (particularly in the case of sedimentary rocks), and minor variations in water content may evoke extreme variations in dielectric constant.

Dielectric Constant of Anisotropic Rocks

An orderly arrangment of grains according to form and crystallographic axes, as well as alternating layers of rocks with differing mineral composition, may cause anisotropy. In discussing physical properties of rocks, we may distinguish between microanisotropy and macroanisotropy. Microanisotropy may arise when there is a preferred orientation of grains comprising igneous, metamorphic, or sedimentary rocks. In sandstones, shales, and other sedimentary rocks, mineral grains which are platey or elongate may settle out with their long dimensions in the plane of sedimentation causing anisotropy in physical properties. Microanisotropy may be a primary characteristic of a rock, or it may arise during later dynamic metamorphism. In many cases, orientation by external grain form may reflect orientation by internal structure. In such cases, the anisotropy will be larger for minerals which have the largest variation between dielectric constants measured in the principal directions ε_1, ε_2, and ε_3. Interbedding of layers with different mineral compositions leads to macroanisotropy.

The dielectric constant of a layered rock depends on the direction, relative to the bedding planes, in which it is measured. Let us consider the case in which a rock is made up of layers of two different materials and the external electric field is applied parallel to the layering. The equivalent circuit for such a dielectric consists of two lossless capacitors, C_A and C_B, connected in parallel, with shunting resistances, r_A and r_B (see Fig. 9). The voltage drop across each of the layers is the same. The dielectric constant in this case is

$$\varepsilon = \Theta_A \varepsilon_A + \Theta_B \varepsilon_B, \qquad\qquad (II.3)$$

where ε_A and ε_B are the dielectric constants for the two components, and Θ_A and Θ_B are the volume fractions of each of the components.

For the case in which the external field is applied in a direction perpendicular to the bedding planes, the equivalent circuit consists of the series connection of the capacitors C_A and C_B and resistances r_A and r_B. In this case, the approximate expression for the dielectric constant of the composite is

$$\varepsilon = \frac{\varepsilon_A \cdot \varepsilon_B}{\Theta_A \varepsilon_A + \Theta_B \varepsilon_B} .$$ (II.4)

Thus, knowing the properties of the components and the direction of the field relative to the layering, it is possible to calculate the dielectric constant for a layered material.

Laboratory studies of the variation of dielectric constant with direction have been carried out by Rao [41] and Stacey [42]. Rao has described measurements of the dielectric constant made in three mutually perpendicular directions. To do this, three samples were cut from each rock, one of which was perpendicular to the direction of the rock fabric, and two of which were parallel to the fabric but perpendicular to each other. The samples were dried in an oven at 200°C for 4 hr. Measurements were also made with the samples artificially resaturated with water. In the water-bearing samples, the value for ε depended more strongly on direction that those for dry rocks. For the later, as seen from the data listed in Table 6, the values for ε measured in the three directions are nearly the same. Exceptions to this are the samples

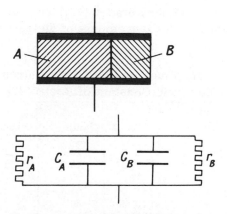

Fig. 9. Equivalent circuit for a capacitor consisting of dielectrics connected in parallel.

Table 6. Anisotropy of the Dielectric Constant of Rocks

Rock	$\varepsilon_{1(\perp)}$		$\varepsilon_{2(\parallel)}$		$\varepsilon_{3(\parallel)}$	
	dry	wet	dry	wet	dry	wet
White marble	7.7	7.9	7.7	7.9		
Gray marble	6.2	6.8	8.3	9.4	8.3	9.0
Marble	8.5	8.7	8.7	8.8		
Limestone	8.3	10.4	8.5	12.2	8.4	12.4
"	8.6	10.2	8.5	12.5	8.6	12.2
"	7.8	10.6	7.8	12.0	7.9	12.1
Keratophyre	5.7	5.9	5.5	5.9		
Norite	10.9	11.3	12.0	12.7		
Chloritized dolerite	8.3	9.7	8.2	10.7		10.7
Leptite	6.1	6.4	6.1	7.1	6.3	7.1
Deccan trap rock	11.2	13.0	12.0	15.5		
Eclogite	12.7	17.0	9.7	14.1	9.3	13.6
Granite	5.9	8.4	5.9	8.7	5.8	8.7
"	7.0	7.1	7.0	7.2	6.8	7.3

of eclogite, norite, and trap rock from the Deccan. Anisotropy in these rocks appears to arise from preferential orientation of rutile or other heavy minerals. Also, there is a marked anisotropy in the case of gray marble, both in the dry and moist conditions. In this case, it is obvious that there is a preferred orientation of calcite grains which reflects internal atomic structure. In all the anisotropic rocks, including eclogite, the maximum value for the dielectric constant is observed in the wet state.

In the rocks studied by Stacey [42], an even more pronounced anisotropy was found. Measurements of dielectric constant made

Table 7. Maximum and Minimum Values for the
Dielectric Constant of Rocks [42]

Rock	$\varepsilon_{max\parallel}$	$\varepsilon_{min\perp}$	$\varepsilon_{max}/\varepsilon_{min}$
Adamellite	15.9	12.5	1.27
"	29.37	18.82	1.56
Granite	107	25.5	4.19
Paragneiss	101.6	65.4	1.55
Gneiss	116.66	80	1.45
Rhyolite	1000	300	3.0
	233	208	1.1
Pegmatite	41	36	1.13

in two mutually perpendicular directions, one along the layering
and one across the layering, are listed in Table 7, summarized in
the form of arithmetic averages from the originally published data.
In this work, the direction of layering was defined in terms of the
preferred orientation of grains of magnetic minerals. The extremely high values for the dielectric constant listed here may possibly be explained by the effect of conducting minerals contained in
the rocks, or possibly by water content, inasmuch as the measurements were made on samples at a relatively low frequency, 1592
cps. In these rocks, as one would expect, the maximum value
for the dielectric constant was observed for a direction parallel to
the layering. The angle φ between the maximum value for ε and
the direction of the layering planes generally is no more than $\pm 5°$,
except for paragneiss ($\varphi = \pm 10°$) and rhyolite ($\varphi = \pm 20°$). The most
extreme anisotropy was found in the rocks with the highest values
for the dielectric constant, as, for example, granite and one sample
of rhyolite. The highest values for the dielectric constant observed
along the layering direction, in comparision with those observed
in a direction normal to the layering, are in agreement with
equations (II.3) and (II.4), which require just such a relationship.

The various data in Tables 6 and 7 indicate that anisotropy
in the dielectric constant in rocks may arise from structural peculiarities. Therefore, in studing the dielectric constant of rocks,
information may be obtained which is of value in petrography and
geophysics.

Effect of Pressure on the Dielectric Constant of Rocks

The effect of pressure on the dielectric constant of rocks
appears to be the least studied of the parameters we have considered. Most of the research on this phenomenon has been done
with gases and liquids. A limited amount of work done with solid
materials suggests that the dielectric constant decreases with
increasing pressure. Mayburg [43], in working with the ionic
crystals MgO, LiF, NaCl, KCl, and KBr, found that an increase
in pressure from 0 to 800 kg/cm^2 decreased the dielectric constant of MgO by 2.56% and KBr by 9.32%. This change results
not only from a reduction in volume of a crystal but also from a
reduction in the internal field. Similar results have been reported

for polyethylene plastic, for silica glass, and for diamond [44, 45, 46]. Contradictory results have been reported for measurements on barium titanate. Vul and Vereshchagin [47] have found an increase in the dielectric constant with increasing pressure, but Kozlobaev [48] has found a decrease.

The variation in dielectric constant with pressure for a material in which polarization is by electron or ion displacement may be predicted using the Clausius–Mosotti equation.

First let us consider the simplest case, in which polarization takes place only by electron displacement. The change in dielectric constant in such a solid material can be caused only by a change in the number of polarizing particles.

Differentiating the Clausius–Mosotti equation with respect to pressure and holding the temperature constant

$$\frac{d}{dp}\left(\frac{\varepsilon-1}{\varepsilon+2}\right) = \frac{4}{3}\pi n\alpha_e \frac{1}{n}\frac{dn}{dp} \; ;$$

therefore,

$$\frac{1}{\varepsilon}\frac{d\varepsilon}{dp} = \frac{(\varepsilon-1)(\varepsilon+2)}{3\varepsilon}\cdot\frac{1}{n}\cdot\frac{dn}{dp} \, , \qquad\qquad (II.5)$$

where $(1/n)\,(dn/dp)$ may be taken as the coefficient for volume compressibility, β_V. This allows us to write equation (II.5) in the following form:

$$\frac{1}{\varepsilon}\cdot\frac{d\varepsilon}{dp} = \frac{(\varepsilon-1)(\varepsilon+2)}{3\varepsilon}\,\beta_V.$$

Thus, for constant temperatures, the variation in the dielectric constant with pressure is a function only of the coefficient for volume compressibility.

For dielectrics in which polarization by ion displacement takes place in addition to polarization by electron displacement, the dependence of the dielectric constant is determined both by the compressibility and by the decrease in ionic polarizability. In this case, we differentiate the second form of the Clausius–Mosotti equation with respect to pressure:

$$\frac{\varepsilon-1}{\varepsilon+2} = \frac{4}{3}\pi n\,(\alpha_1+\alpha_2+\alpha_i),$$

where α_1 and α_2 are the electronic polarizabilities of the ions, α_i is the polarizability by ion displacement, and n is the number of ions per cubic centimeter.

After differentiating with respect to pressure, we have

$$\frac{d}{dp}\left(\frac{\varepsilon - 1}{\varepsilon + 2}\right) = \frac{4}{3}\pi\left(\alpha_1 + \alpha_2 + \alpha_i\right)\frac{dn}{dp} + \frac{4}{3}\pi n\frac{d\alpha_i}{dp}.$$

But $(1/n)\,(dn/dp)$, which is the coefficient for volume compressibility of a solid, is very nearly three times the linear coefficient of compressibility β_l, so that we may write

$$\frac{3}{(\varepsilon + 2)^2}\frac{d\varepsilon}{dp} = \frac{4\pi}{3}n\left(\alpha_1 + \alpha_2 + \alpha_i\right)3\beta_l + \frac{4}{3}\pi n\frac{d\alpha_i}{dp}.$$

Remembering that

$$\frac{4}{3}\pi n\left(\alpha_1 + \alpha_2 + \alpha_i\right) = \frac{\varepsilon - 1}{\varepsilon + 2},$$

we obtain

$$\frac{1}{\varepsilon}\frac{d\varepsilon}{dp} = \left[\frac{\varepsilon - 1}{\varepsilon + 2}3\beta_l + \frac{4\pi n}{3}\frac{d\alpha_i}{dp}\right]\frac{(\varepsilon + 2)^2}{3\varepsilon}. \tag{II.6}$$

In order to calculate $(1/\varepsilon)\,(d\varepsilon/dp)$, it is necessary to know the derivative of ionic polarizability with respect to pressure, $d\alpha_i/dp$, which may be determined as follows. Let us make the assumption that at high pressures, where an increase in density can take place only by a decrease in interatomic distances, polarizability changes only because of changes in atomic spacing. In this case, the binding energy may be expressed by a simple power law. The elasticity coefficient, K, may be computed with the formula:

$$K = \frac{2\alpha_M\,(n_1 - 1)\,q^2 r_0^{n_1 - 1}}{3 r^{n_1 + 2}},$$

where α_M is a numerical parameter (the Madelung constant, which assumes the value $\alpha_M = 1.74$ for an NaCl-type lattice), q is the charge of an ion, r_0 is the least spacing between ions, r is the spacing between ions, and n_1 is a coefficient determined from the mechanical properties of a crystal.

For halite, the values of n_1, reported by various authors, range from 7.8 to 11.3. For other salts, the value for n_1 falls in the range from 3.96 to 19.7.

Differentiating this last equation with respect to pressure, we have

$$\frac{dK}{dp} = -\frac{2\alpha_M (n_1-1)(n_1+2) q^2 r_0^{n_1-1}}{3 r^{n_1+3}} \cdot \frac{dr}{dp} = -(n_1+2) K \frac{1}{r} \frac{dr}{dp}.$$

However, $(1/r)(dr/dp) = \beta_l$, the coefficient of linear compressibility, and therefore

$$\frac{dK}{dp} = (n_1+2) K \beta_l.$$

Considering that the polarizability for ion displacement is $\alpha_i = 2q^2/K$, we have

$$\frac{d\alpha_i}{dp} = \frac{2q^2}{K^2} \frac{dK}{dp} = \frac{\alpha_i}{K} \frac{dK}{dp}$$

or

$$\frac{d\alpha_i}{dp} = \alpha_i (n_1+2) \beta_l. \tag{II.7}$$

Substituting the expression (II.7) into equation (II.6) we have

$$\frac{1}{\varepsilon} \frac{d\varepsilon}{dp} = \left[3 \frac{\varepsilon-1}{\varepsilon+2} \beta_l + \frac{4\pi n}{3} \alpha_i (n_1+2) \beta_l \right] \frac{(\varepsilon+2)^2}{3\varepsilon}. \tag{II.8}$$

Using equation (II.8), the pressure coefficient for the dielectric constant for a solid exhibiting electron and ion displacement polarization may be calculated theoretically to a first approximation.

In contrast with other materials, an increase in dielectric constant with increasing pressure is observed for rocks [29, 49, 50, 51]. For application of uniaxial pressure, the largest change in dielectric constant is observed for pressures in the range 0 to 600 kg/cm^2. Data for the ratio of the dielectric constant at a pressure p to the dielectric constant of the same sample with no pressure applied, $\varepsilon_p/\varepsilon_0$, are shown graphically in Fig. 10. The general character of the change in dielectric constant with pressure is the same for all of the rocks, but the magnitude of change

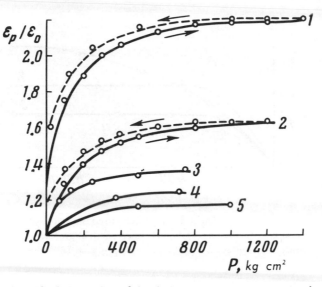

Fig. 10. Relationship of the dielectric constant ε_p measured at a pressure p and the dielectric constant ε_0 measured at atmospheric pressure at uniaxial pressure p: (1) sandstone, (2) syenite, (3-5) granite. Arrows indicate direction of pressure change.

varies. The most radical changes in the ratio $\varepsilon_p/\varepsilon_0$ were observed for sandstone and syenite; a smaller variation was observed for several granites.

For pressures over 600 kg/cm^2, the rate of increase in $\varepsilon_p/\varepsilon_0$ diminishes, and the ratio approaches a limit.

It has been suggested that the extreme variation of ε at low pressures, in the range up to 100 or 200 kg/cm^2, may be explained by an improved contact between the electrodes used in making a measurement and the surfaces of a sample. Reduction in the series capacitance caused by an air gap between the electrodes and the sample leads to a larger value for the capacitance of the sample and a correspondingly larger value for the dielectric constant. Apparently, this is the explanation of the large increases in dielectric constant observed by Rudajev [49] in the pressure range 0 to 50 kg/cm^2.

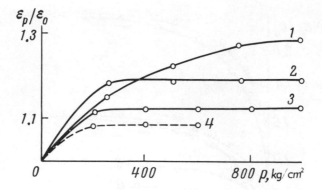

Fig. 11. Relative change in the dielectric constant for hydrostatic pressures up to 1000 kg/cm^2: (1) limestone, (2) granite, (3) basalt, (4) diabase.

As indicated in this last reference, the values observed for ε as pressure is relieved do not differ significantly from values observed for ε as pressure is applied, providing no permanent deformation of the sample is caused by the application of pressure. The values for dielectric constant observed during reduction of pressure are somewhat larger than values observed during application of pressure.

When measurements are made with triaxial pressure being applied, the results are very nearly the same as those obtained with uniaxial pressure. The curves shown in Fig. 11 indicate a rapid change in ε with pressure at low pressures. However, according to reference [29], the rate of change of ε with triaxial pressure is less than the rate of change with uniaxial pressure. This may be explained by the failure to make corrections for the reduction in dimensions of a sample under uniaxial pressure; therefore, the triaxial measurements are probably reliable.

The dielectric constant of rocks continues to increase with pressure up to pressures of 50,000 kg/cm^2, according to Bondarenko [50]. The pressure coefficient for the ratio of dielectric constants

$$\frac{1}{\varepsilon_p / \varepsilon_0} \frac{d\,(\varepsilon_p / \varepsilon_0)}{dp}$$

for the rock dunite is 6×10^{-6} per kg/cm^2 over the pressure range up to 20,000 kg/cm^2 and is somewhat less in the range from 20,000 to 40,000 kg/cm^2 (about 2.5×10^{-6}). The dielectric constant for serpentinite, porphyry, and basalt increases more rapidly with pressure.

The increase in the dielectric constant of rocks with increasing pressure may be explained by an increase in density by the closing of pore structures. It should be noted that the increase in the dielectric constant is compatible with variations in acoustic velocity, Young's modulus, and compressibility as a function of pressure [52].

Studies of the relationship of dielectric constant to pressure are becoming increasingly important as electrical prospecting is carried to greater and greater depths. Moreover, the study of values for ε at high pressures and temperatures provides us with an additional tool for investigating the interior of the earth.

In the last chapter, the relationship between dielectric constant and temperature and frequency is considered along with a discussion of dielectric loss, inasmuch as the two subjects are closely related.

Chapter III

Electrical Resistivity of Rocks

Solid materials may be divided into three groups on the basis of the magnitude and mechanism of electrical conduction – conductors, semiconductors, and dielectrics (solid electrolytes). The highest conductivities are found in metals and their alloys, with values of conductivity ranging from 10^3 to 10^6 mho/cm.

In contrast to these materials are the solid electroltes, which generally have conductivities less than 10^{-9} mho/cm. Between these two extreme types of material lie the semiconductors, with conductivities in the range from 10^4 to 10^{-9} mho/cm. It should be noted that there is no sharp boundary between conductors and semiconductors on one hand, or between semiconductors and insulators on the other hand. The distinction between classes is based not so much on the value of conductivity as on the mechanism.

Electrical conductivity in metals and semiconductors is provided by the same charge carriers – electrons. The basic difference between these materials is the energy required for activation of the electrons. The activation energy for electrons in a metal is zero, while in semiconductors, it ranges from a few tenths to several electron volts. The two types of conductors also vary in their temperature dependence. In metals, the maximum conductivity is found at temperatures close to absolute zero, while semiconductors are nearly insulators at low temperatures.

The very much lower values of conductivity observed for semiconductors as compared with those for metals are explained by the fact that only a small fraction of the valence electrons take part in conduction.

Charge carriers in semiconductors and insulators are activated in the same manner, which leads to some similarities between the two groups of materials. However, the fact that the charge carriers in insulators are ions rather than electrons leads to significant differences between the two groups of materials. The essential difference is in the mobility of the charge carriers. Electrons may move rather freely over the energy barriers between atoms, while ions have much larger barriers to surmount in moving from one lattice position to another.

Rocks, and the minerals comprising them, belong to one or another of three classes of conductors. The conductivity of rocks varies over much wider limits than any of the other physical properties. Some minerals may be classed as metallic conductors, while others have high enough resistivities to be classed as insulators. Many ore minerals and ore-bearing rocks exhibit semiconductor properties. However, the great majority of minerals are insulators. Therefore, we must review briefly the principles of conduction, particularly in dielectric materials.

Brief Review of the Electrical Conductivity of Dielectrics

As is well known, in a dielectric material, the atoms, molecules, and ions are not free to move over large distances. However, any dielectric material will contain some amount of weakly bonded charged particles. When an external field is applied, these weakly bonded charges are set in motion to form an electrical current. The current density, j, depends on the number and velocity of charge carriers per unit volume according to the equation

$$j = nqv,$$

where n is the number of charge carriers per cm^3, q is the charge on each particle, and v is the velocity of the particles.

The velocity of the charge carriers is proportional to the applied electric field E

$$j = nquE, \qquad (III.1)$$

where u is the mobility of the charge carriers.

Current in a dielectric material may be carried not only by ions but also by electrons [12]. Materials are known which exhibit both electronic and ionic semiconduction, either simultaneously or separately in different temperature ranges. For example, copper iodide, CuI, is an electronic semiconductor at low temperatures. At a temperature of 250°, conduction by ions becomes as important as electron conduction, and above 400°, electrical conduction is almost entirely by ion motion [53]. Quartz and feldspar provide examples of materials in which the nature of conductivity depends on field strength. In weak fields conduction is by ions, while in strong fields conduction is by electrons [54].

If current flow is contributed by several species of charge carriers (as for example, cations, anions, and electrons), the current is given by the formula

$$j = (n_1 q_1 u_1 + n_2 q_2 u_2 + \ldots + n_i q_i u_i) E.$$

It should be observed that materials in which both cations and anions contribute simultaneously to current flow are quite rare [12]. The relative abundance of a particular ion as a charge carrier depends on its valance and size. Current is carried preferentially by ions of a given size which have the largest charge, or by ions of a given valency which have the smallest size.

The nature of electrical conductivity in a particular material is established experimentally. Criteria for distinguishing the various conduction mechanisms are Faraday's law and the Hall effect, as well as thermoelectrodynamic forces. Electronic conduction can be verified by the existence of the Hall effect or by thermoelectrodynamic forces.

The Hall effect is characterized by the development of a transverse voltage when a magnetic field is applied perpendicular to the direction of current flow in a material. It is proportional to the current, I, and the magnetic field strength, H, and inversely proportional to the thickness of the sample, d, so that

$$V = R \frac{IH}{d},$$

where R is the Hall coefficient, which is a function of the number of charge carriers, n, and the charge on each, q. For electronic conduction R is negative, while for hole conduction it is positive. Knowing the Hall coefficient, it is a simple matter to compute the velocity and number of charge carriers using the formula

$$R = \frac{1}{nq} \text{ and } \sigma = nqu. \tag{III.2}$$

It has been verified, using Faraday's law, that conduction in the majority of solid dielectrics is ionic. At high temperatures, it is possible to use Faraday's law, equating the charge transferred through a material to the quantity of metal precipitated at the electrodes.

Since ionic conduction is so common, we must review the mechanisms of such conductions in more detail.

Ionic Conductivity in Dielectrics. In a real crystal, in addition to the ions in the proper lattice positions there are also impurity ions and ions which represent lattice defects. Therefore, depending on which type of ion is contributing to conduction, ionic-conducting crystals may be divided into two classes – impurity and defect conductors. Defect conductors are characterized by motion of ions from the crystal lattice, which are most abundant at high temperatures. Inasmuch as impurities vary from sample to sample and may have a variety of chemical properties, conductivities for crystals may show wide scatter [53, 55]. Academician A. F. Ioffe and his coworkers obtained exceedingly pure crystals of alum by repreated crystallization which had an electrical conductivity three orders of magnitude lower than the normal conductivity of alum. In dielectrics with atomic or molecular lattice structures, as well as in amorphous dielectrics, the absence of ions means that conduction is entirely by impurities.

Unrelated to either of these mechanisms is conduction by normal lattice ions which have been displaced from their lattice

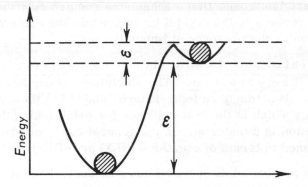

Fig. 12. Variation of the potential energy of an ion
on its transfer to an interstitial position in a crystal
lattice (from Frenkel): lower — normal ion position,
upper — interstitial ion position.

positions by thermal agitation. A small fraction of the lattice ions
will always be displaced to interstitial positions in the lattice (see
Fig. 12). The energy required to displace an ion to an interstitial
position is termed the energy of dissociation, ε. In order for these
ions to contribute to current flow, they must move to neighboring
lattice positions vacated by other ions going into interstitial positions. It is necessary only to provide the energy ω to drive an interstitial ion over the energy barrier into a neighboring lattice
position. An ion may travel from one lattice position to a neighboring lattice position through lattice defects or through lattice
pore spaces which are not adjacent to ions of the same sign. The
application of an external field leads to a preferential direction of
motion for these ions. When no external field is applied, the direction of movement of the interstitial ions will be random. It may
be shown that the effective current provided by interstitial ions is
proportional to the applied field, E.

In addition to the process described above, it is possible
for interstitial ions to move from one interstitial position to
another interstitial position, rather than back into a lattice position. Such motion occurs when the energy barrier between interstitial positions is less than the dissociation energy barrier, which
would have to be overcome to return an ion to a lattice position
[53].

These last conduction mechanisms require only the ions normally present in the crystal lattice, while the earlier mechanisms require the presence of impurity ions.

Relation of Conductivity in Dielectric Crystals to Temperature. The electrical conductivity of dielectrics depends strongly on temperature, and it is this temperature dependence which is the best evidence for establishing the nature of conduction in a material. In the general case, conductivity may be expressed in terms of equations (III.1) and (III.2).

The number of dissociated ions, n, depends on temperature according to the equation

$$n_t = n_0' e^{-\frac{\mathscr{E}}{kT}},$$

where n'_0 is the total number of ions per cm^3, $\mathscr{E}$ is the dissociation energy, and kT is the thermal energy.

Ionic mobility is also an exponential function of temperature

$$u_t = u_n e^{-\frac{\omega}{kT}}, \tag{III.3}$$

where u_n is the limiting mobility of an ion, and ω is the energy required to displace an ion. Substituting these expressions for n_t and u_t in equation (III.2) and noting that n'_0, u_n, and q can be replaced with the single parameter σ_0, we have

$$\sigma_t = \sigma_0 e^{-\frac{E_0}{kT}}, \tag{III.4}$$

where $E_0 = \mathscr{E} + \omega$ is the activation energy for ions contributing to current flow.

In practice, temperatures are usually expressed in the centigrade rather than the absolute scale. In this case, electrical conductivity may be computed using the approximate expression

$$\sigma_t = \sigma_0 e^{-2\alpha t},$$

where σ_0 is the conductivity at $t = 0°C$ and α is the temperature coefficient.

For crystals in which only one species of ions contributes to conduction, the expression for $\sigma_t = f(1/T)$ may be rewritten as

$$\ln \sigma_t = -\frac{E_0}{kT} \ln \sigma_0$$

or in logarithms

$$\log \sigma_t = \log \sigma_0 - 0.43 \frac{E_0}{kT} .$$

A graphical plot of this equation has the form of a straight line with a slope which is proportional to the activation energy for the charge carriers.

If $1000/T$ is plotted along the abscissa and E_0 is expressed in electron volts, then the slope angle, φ, is given by

$$E_0 = 0.2 \tan \varphi. \tag{III.5}$$

The relationship between conductivity and temperature has been determined to be linear in this manner over some temperature interval for a number of minerals (quartz, muscovite, periclase, calcite). The activation energy for the charge carriers in quartz was found to be 0.88 eV along the optical axis, and 1.07 to 1.32 eV normal to the optical axis. For periclase, the activation energy was found to be 1.16 eV and for mica, 0.75 eV [12].

The relationship between conductivity and temperature may be linearized in this manner only in those cases in which there is but one species of ion contributing to conduction, or when there is only one value for activation energy. Published data indicate that, at higher temperatures, in many materials the type of charge carrier changes. At high temperatures, charge carriers which require very high activation energies become important. Thus, the slope of the curve changes, becoming flatter [12, 56, 57]. A particularly notable change in the character of conductivity in an ionic crystal with impurities in going from low to high temperatures is shown by the example in Fig. 13. The change in slope of this curve indicates that the type of charge changes as the temperature is increased. Usually, impurity ions provide the charge carriers. The binding energy of impurity ions is somewhat less than the binding energy of regular lattice ions and, as a result, they are

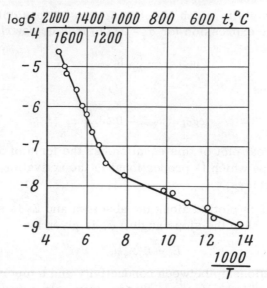

Fig. 13. Dependence of the resistivity of alundum on temperature.

available for conduction at lower temperatures. Moreover, the freeing of impurity ions makes room for regular lattice ions to move later. If a phase change takes place, there is an abrupt discontinuity in the $\log \sigma = f(1/T)$ curve.

The expression for conductivity, considering both impurity and lattice ions, contains two terms

$$\sigma_t = \sigma_1 e^{-\frac{E_0'}{kT_1}} + \sigma_2 e^{-\frac{E_0''}{kT_2}}.$$

The first term corresponds to low temperature conductivity, while the second corresponds to high temperature conductivity.

All of these considerations, which are based on conduction in a single crystal, may be extended to a polycrystalline material.

In measuring the temperature dependence of electrical conductivity in a dielectric crystal, one should keep in mind the possibility of irreversible reactions, either mechanical or chemical, taking place at high temperatures.

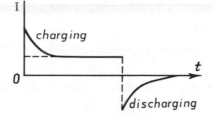

Fig. 14. Relationship between current
I and time t for application of an electric field (charging) and removal (dis-
charging) to a solid dielectric.

Appearance of Polarization Accompaning Current Through a Solid Dielectric. Electrically conducting dielectrics are characterized not only by ion flow, but in many
cases also by a significant variation of current with time at low
temperatures, related to polarization [12, 53, 58, 59]. This phenomenon, which occurs following the flow of direct current, is
widely used in geophysical exploration. Therefore, it is worth our
while to consider this effect more carefully.

Shortly after a current begins to flow, the amount of current
decreases to some constant value. When the current circuit is
broken, a transient current, decaying to zero, is observed
(Fig. 14). The observed change of current with time is compatible
with the collapse of an electric field within the dielectric, with the
rate of collapse depending on the process controlling it. It is
known that dielectric polarization decays very rapidly (in hundredths of a second). In contrast, the decay of volume polarization
(high-voltage polarization) takes place much more slowly, re-
quiring hours or even days.

Skanavi [12] has listed the following mechanisms for storing
charge in a solid dielectric.

1. Interfacial polarization in an inhomogeneous dielectric,
   such as layered material, with charge accumulating at
   boundaries between regions.

2. Development of volume charge, distributed throughout the material.

3. Collection of impurity ions in surface layers at the electrodes.

4. Formation of a conductive region in the sample around one or both electrodes.

Depending on the length of time current is driven through a sample, the third and then the fourth mechanism may take place in the same sample.

In rocks, it should be noted that polarization may also be related to oxidation – reduction processes which form an ionic double layer at boundaries between solid and liquid phases as well as to other phenomena which are less important.

In order to determine which of these processes cause the flow of direct current to decrease with time, it is essential to know the distribution of electric field within the dielectric, which may be accomplished, for example, using the probing method suggested by Ioffe. This method consists of measuring the voltage between pairs of electrodes located at equal intervals along a sample. This probing method allows the principal mechanism of charge accumulation to be determined when an adequate number of samples are used. Figure 15 shows how electric field varies within a sample for the different forms of charge storage.

Ioffe found clear evidence of the storage of charge in a thin layer about a micron thick at the cathode for crystals of calcite and potassium nitrate and charges of both polarities distributed through a region with appreciable thickness close to the anode and cathode in the case of quartz [53]. The accumulation of charge in a dielectric develops an electric field which is opposed to the external field. Therefore, polarization, P, developed by current flow reduces the amount of current flow through a crystal.

In contrast to liquids, in which the emf caused by polarization may be as large as several volts, in a solid dielectric it may be as large as several thousand volts and, in many cases, may be nearly equal to the applied voltage. Therefore, this type of polarization is termed high-voltage polarization.

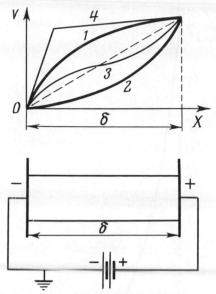

Fig. 15. Distribution of a potential V
through a sample: (1) for a positive
volume charge, (2) for a negative volume
charge, (3) for volume changes of both polarities, (4) with the formation of filaments;
δ is the length of the sample (from Ioffe).

Impurities play a very important role in the development of
high voltage polarization. The accumulation of charge is less in
pure crystals. This has been well established by the work of
Gokhberg and others [60-62] and confirmed by data from Icelandic
and American calcite. In the American calcite, which was the
purer of the two varieties, high voltage polarization was almost
completely absent. Thus, the cleaner the material, the lower
will be the high-voltage polarization.

The development of this polarization is a relatively slow
process which depends on the physical and chemical properties of
the material. At the instant current flow starts, $P = 0$ and the
potential drop through the sample is linear ($dv/dx = const$). As
current flow continues, charge accumlates and the potential drop
changes until the maximum value for high voltage polarization,
P_{max}, is reached [57].

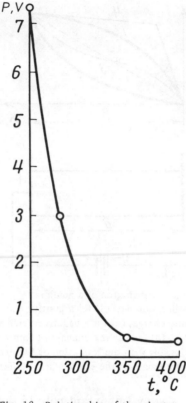

Fig. 16. Relationship of the electro-
motive force of high-voltage polarization P to temperature for a particular sample of rock salt (halite):
charging time $- 2.5 \cdot 10^{-2}$ sec, charging voltage $- 300$ V.

The maximum value of polarization depends not only on the
duration of current flow, but also on temperature and the applied
electric field. At high temperatures, diffusion and recombination
take place more rapidly, decreasing the emf caused by polarization, according to data given by Alexandrov, nearly to zero. Very
nearly identical behaviors were found for a number of dielectric
materials with different compositions and structures [57]. In a
very recent paper [63], experimental work indicated that for halite,
polarization emf was much reduced at high temperatures but did

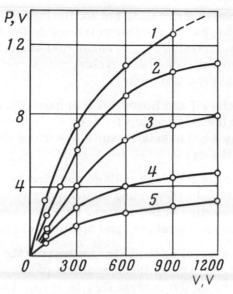

Fig. 17. Relationship of the emf of high-
voltage polarization in halite to the charging voltage at various temperatures: (1)
128°, (2) 216°, (3) 238°, (4) 251°, (5) 290°C.

not disappear completely. It appears to degenerate into low-volt-
age polarization, which approaches a constant value at some temperature (Fig. 16). It is possible that chemical composition and
physical properties are dominant factors in determining the
character of change in high voltage polarization with temperature,
and therefore there may be different types of relationships $P = f(t)$
for different materials.

Along with the decrease in P_{max} at high temperatures,
a shortening of the time required for polarization to take place is
observed. At sufficiently high temperatures, the time required
for polarization to take place becomes insignificant.

For weak fields, the polarization emf increases linearly with
the applied field strength. In strong fields, the polarization
reaches a maximum and then remains constant or decreases as the
external field is increased. Figure 17 shows some graphical relationships $P = f(V)$ for halite at different temperatures, observed
by Kosman and Petrova [63]. All the curves have the same form,

nearly linear close to the origin, but exhibiting saturation at higher applied field strengths. A similar relationship for $P = f(V)$ has been reported by Aleksandrov. Presnov [64] has noted a decrease in polarization emf at even higher fields, measured in kilovolts (from 2 to 1000 kV/cm, and higher).

Venderovich [65] and Lozovskii [66] have discussed the theoretical basis for the relationship $P = f(V)$. In one or the other of these papers, the experimental results described above have been substantiated in theory.

Thus, the amount of high voltage polarization may change from sample to sample, depending on the quantity and the physical and chemical properties of impurities, as well as on external factors such as time, temperature, and applied field strength.

There are several methods for measuring the emf from high-voltage polarization. One of these consists of selecting an applied voltage, V_i, for which the current flow is zero. Under this condition, we have $V_i = P$. Another method consists of measuring current flow for two different levels of applied voltage, V_1 and V_2, with which are associated a steady current, I_1, and a suddenly increased current, I_2. With these data, the value for P may be determined with the formula

$$P = \frac{I_1 V_2 - I_2 V_1}{I_1 - I_2}.$$

It should be noted that these methods permit only approximate determinations of P because of the finite response time of measuring equipment and the time required to make a reading. Other methods have been described in the literature [67–69].

Inasmuch as current flow varies with time in many dielectrics, it is useful to define initial, terminal, and "true" values of resistance [12, 70, 71]. The initial resistance is that met by the current when it first begins to flow. Skanavi [12] indicates that the "true" initial resistance can be measured only with a decreasing current, so that polarization is confined to a thin surface layer. The terminal resistance is calculated from the amount of current flowing after it has dropped to a stable or terminal value.

The most reliable measure of the conduction characteristic of a dielectric material is the "true" resistance. This is calculated after subtracting the polarization emf from the applied voltage.

As the polarization emf in a dielectric varies from zero to its maximum value, the internal electric field also varies. At a given instant, it is given not in terms of the applied voltage, V, but in terms of the voltage difference, $V - P$. Therefore, Ohm's law, according to Ioffe, should have the following form for a polarized dielectric:

$$I = \frac{V - P}{R_{true}}.$$
(III.6)

Using this expression, we may compute the true resistance.

It should be kept in mind that not only is there a quantitative difference between the true and terminal values of resistance, but also the relationship $R_{true} = f(V, t)$ has a different character than the relationship $R_{term} = f(V, t)$. The terminal resistance is more a function of temperature than is the true resistance because of the effect of diffusion on polarization at higher temperatures. At high temperatures, the terminal resistance is close to the true resistance, and at some temperature may even be equal to it. The true resistance for a material poor in impurities does not depend on field strength in weak fields and may remain constant up to very high field strengths [62]. The terminal resistance decreases with increasing field strength, inasmuch as the polarization emf is not proportional to the applied field, but approaches saturation for some specific value of E. Moreover, Ohm's law may not be valid at high field strengths because of the addition of electrons to current flow [54, 72].

Surface Conduction in Solid Dielectrics. Surface conductivity, as well as volume conductivity, may be observed in dielectric materials. This is a function of the adsorption characteristics of a material. The formation of an adsorbed layer of moisture depends primarily on the physicochemical properties of a material and on the character of the surface. Polar, porous materials with large specific surface areas or dirty surfaces may exhibit very appreciable amounts of surface conduction. In such materials, the conductivity depends strongly on atmospheric humidity. Such factors are not important in surface conduction in hydrophobic dielectrics, that is, dielectric materials which are not wet by water.

Electrical Conductivity of Semiconductors

The majority of ore minerals – oxides, tellurides, sulfides, and selenides – are semiconductors. In addition to these chemical

compounds, a number of elements, located between the metal and
the insulators in the periodic table, exhibit semiconductors properties. A portion of Mendeleev's periodic table of the elements is
given in Table 8, which lists 12 elements that have semiconductor
properties. To the right of each chemical symbol is listed the
activation energy which is characteristic of a given semiconductor.
This is the amount of energy required to cause an electron to serve
as a charge carrier.

The systematic variation of activation energy both horizontally
and vertically in the periodic table is indicative of the association
between the electrical properties of semiconductors and their positions in the periodic table. It should be noted that in some cases,
semiconductor properties are exhibited by these elements only under special conditions. For example, gray tin is a semiconductor
at temperatures below 13°C. Some elements which are insulators
under normal conditions (red phosphorus, sulfur, iodine, and gray
selenium) become semiconducting under the influence of light. Carbon in the form of diamond is an insulator, and another form of carbon – graphite – exhibits semiconductor properties in one crystallographic direction and metallic properties in another.

Experimental studies of the magnitude and sign of the Hall
effect in semiconductors indicate that the Hall coefficient R not only
varies in magnitude but also changes sign. This suggests that
charge carriers may carry either positive or negative charges.
On the other hand experimental data such as the absence of electrolysis products in semiconductors and the tremendous amount

Table 8. Elements with Semiconductors Properties

Period	Group						
	II	III	IV	V	VI	VII	
II	Be	B 1.1	C 5.2	N	O		
III		Al	Si 1.1	P 1.5	S 2.5	Cl	
IV		Ca	Ge 0.75	As 1.2	Se 1.7	Br	
V		In	Sn 0.08	Sb 0.12	Te 0.36	I 1.25	Xe
VI			Pb	Bi	Po	At	

of energy required to provide positions for conduction preclude the possible existence of charge carriers bearing a positive charge.

The existence of two types of conduction may be explained as follows, without resorting to the highly complicated mathematical arguments, by taking two propositions from quantum mechanics [73]. The first proposition is that electrons in free atoms are not free to assume any value of energy whatsoever, but are restricted to a set of permissible energy levels. These levels are separated by rather wide energy bands which are not available to the electrons. The second proposition is that there may be no more than two electrons in any one permissible energy level.

In the solid state, atoms are no longer free and isolated from one another, but are bonded together. As a result of this bonding, in place of N similar energy levels associated with N atoms, in a solid, the energy levels merge into an energy range containing these N levels (Fig. 18). This zone is termed a range of permissible energies. Because the similar levels are very close in energy, electrons may be transferred from one to another by an external electric field as well as by thermal activation. The distance be-

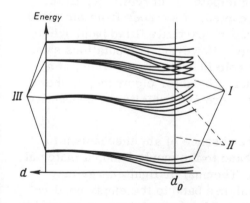

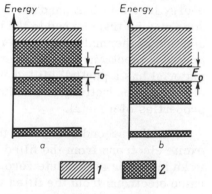

Fig. 18. Conceptual sketch of energy levels in a solid material: d is the distance between atoms in a crystal, d_0 is the least distance between atoms, I is a zone of overlapping energy levels, II is a zone of nearly overlapping energy levels, III is the condition for isolated atoms.

Fig. 19. Two possible conditions for energy bands in a solid: a — metal, b — a non-metal (semiconductor) with $E_0 < 2$-3 eV or a dielectric with $E_0 > 2$-3 eV, where E_0 is the width of the forbidden band between energy bands. (1) Region of free conduction, (2) filled band.

tween zones containing different energy levels may be several electron volts. The transfer of an electron from one energy zone to the next requires activation by this amount of energy. The energy which must be provided equals the distance from one zone to the next. The usual electric field cannot provide this amount of energy. Therefore, interzone transfers of electrons can be excited only by thermal energy.

These are necessary conditions which must be fulfilled for the movement of electrons within or between energy bands, but not sufficient conditions. It is still necessary that no more than two electrons occupy any single energy level, which means that electrons may move only when there are vacant levels to be filled.

There are two possible states for the filling of levels in the two energy bands: (a) one of the bands is partially filled, and (b) one of the bands is completely filled and the other is empty (Fig. 19). In the first case, only enough energy need be added to raise an electron to an empty level in the partially filled band for conduction to take place, and this amount of energy may be obtained from an applied electric field. This leads to current flow with charge carriers bearing negative charges – electrons. The electrons in the filled band do not participate in current flow; all the energy levels in that band are occupied. Electrons from the filled zone may be transferred to the upper, partially filled band by the addition of thermal or light energy. However, computations show that such electrons play a minor role in conduction. Thus, if a material has a partially filled energy band for electrons, it shows electrical conductivity at absolute zero and exhibits the conductive properties of a metal.

In the second case, with the absence of any mechanism to excite electrons from the filled band to the empty band, a material is an insulator at absolute zero. Thermal or light energy may raise electrons from the filled valence band to the empty conduction band. Once in the conduction band, electrons may conduct electricity. When an external field is applied, the electron in the conduction band is raised to a higher energy level because of the energy added by the electric field. In the valence band, a "hole" appears where the conduction electron was removed. Thus, it is possible for an electron in the once-filled valence band to extract energy from the applied field and move into the vacated energy level. Simultaneously, with the motion of an electron in the conduction band, there is motion of an electron in the valence band. According to theory, sufficient movement of electrons take place in the lower band so that the vacant levels are filled. The flow of these electrons under the influence of an external field appears to be a flow of the vacant levels (holes) in the opposite direction. Therefore, in a semiconductor the charge carriers are both electrons and holes. Depending on whether more current is carried by the electrons or by the holes, a semiconductor is termed p-type (primarily hole conduction) or n-type (primarily electron conduction).

The conductivity of a semiconductor is highly sensitive to even minor variations in chemical composition. Impurities serve as sources of charge carriers in semiconductors just as they do in dielectrics. Therefore, semiconductors may be classified in terms of inherent properties and properties related to the presence of impurities. Under impurities are included not only foreign atoms in the crystal structure, but also an excess or deficiency of one of the atoms in the chemical compound as well as various defects in the crystal lattice. Impurity atoms tend to be isolated from one another; therefore, the electron energy levels associated with these atoms do not merge into energy bands, and their discrete energy levels may be found in the zone between the valence and conduction bands of the normal material. If the impurity levels are close to the conduction band, then obviously the transfer of its electrons to the conduction band requires only a little energy. Holes in energy levels in the impurity atoms may not move, depending on the distance to lower energy levels, and therefore not participate in conduction. Thus, conduction is by electrons only. The impurity in this case is termed a donor — it donates electrons.

There is another possiblity: If unfilled levels in the impurity lie just above the valence band for the normal material, hole-type conduction may take place. This happens because of the width of the energy gap between the filled impurity levels and the higher, unfilled conduction band. Therefore, an electron may move readily from the valence band to an impurity level, but cannot move readily from there to the conduction band. The electrons in impurity levels cannot participate in conductions, and this results in purely hole conduction. Impurities which have unfilled levels available to electrons from the valence band are acceptors. If both types of impurity are present in a semiconductor, the sign of conduction in a material will depend on which of the impurities is more abundant.

Electrical conductivity in semiconductors, as well as in dielectrics, is strongly dependent on temperature and may be described with an exponential law

$$\sigma_t = \sigma_0 e^{-\frac{E_0}{2kT}} \ . \tag{III.7}$$

If conduction at low temperatures is dominated by one species of charge carrier and by another species at high temperatures, the relationship $\log \sigma_t = f(1/T)$ consists of two line segments, and the expression for conductivity has the following form:

$$\sigma = \sigma_1 e^{-\frac{E_0'}{2kT_1}} + \sigma_2 e^{-\frac{E_0''}{2kT_2}}.$$

One of the most interesting properties of semiconductors is photoconductivity. In the preceding paragraphs, it was mentioned that electrons might be excited into the conduction band not only by thermal energy but also by light energy. If the energy provided by light is sufficient to raise an electron to the conduction band, it may contribute to electrical conduction. The contribution to conductivity by light-excited electrons or holes is termed photoconductivity – positive and negative. For the majority of photoconductors with positive photoconductivity, the conductivity is linearly proportional to the incident light intensity for weak light and to the square root of the intensity for strong light. In negative photoconductors, light causes a decrease in conductivity.

Equally interesting is the effect of pressure on the conductivity of a semiconductor, which affects the mobility and activation energy of charge carriers, changing the conductivity. The second of these factors is the important one. Theory suggests that activation energy may be increased or decreased when pressure causes the atoms to come closer together. This has also been verified by experiment. For example, the resistivity of n-type germanium is increased by a factor of 4.5 under 30,000 kg/cm^2 pressure, and the resistivity of n-type silicon is reduced by a factor of 2 at the same pressure [73].

At present, experimental data are available about the effect of pressure on the conductivity of semiconductors [74-77] which make it possible to arrive at some significant conclusions concerning the energy spectra of electrons.

Methods for Determining the Resistivity of Rocks

The methods which are used in determining the resistivity of semiconductors and dielectrics in the laboratory may also be used to measure the resistivity of rocks. There are two princpal groups of such methods [78-80]. In one group of methods, direct current is used, and in the other, alternating current. We will consider only the direct current methods in this section. The most commonly used of the direct-current methods are the two-electrode and four-electrode techniques. The primary consideration in using these methods is that of obtaining a minimum amount of contact resistance between the electrodes and the sample. With direct-current methods, very serious consideration should be given to the choice of the proper electrode material. At room temperatures, it is recommended that graphite electrodes in the form of a dispersion be used, either deposited in a thin layer or rubbed onto the sample with a soft pencil. Good results may be obtained with gold or platinum electrodes deposited on the sample in the form of a conductive paint or evaporated onto the samples in a vacuum chamber. Electrical contact is then made to these electrode materials through metal foil or sheets. Sometimes mercury is also used to insure good contact with the rock sample [81]. At higher temperatures (more than 300 to 350°C) graphite electrodes are not satisfactory, inasmuch as graphite burns. Under these conditions, the use of gold or platinum electrodes deposited in a vacuum or electrodes cut from platinum or gold foil [82] is recommended, providing the foil electrodes can be pressed onto an optically flat surface at a pressure of the order of 100 kg/cm^2. The surfaces of the sample must be carefully ground.

Two-Electrode Method. In this method, the amount
of current caused to flow through a prepared sample by a known
voltage is measured. Depending on the resistance of the sample,
various types of current-measuring instruments might be used –
ammeters, milliammeters, microammeters, galvanometers, or
electrometers. The shape of the sample in the two-electrode
method may be that of a cube, a parallelepiped, a cylinder, or a
disc. The sample must have dimensions at least three to five times
greater than the dimensions of the grains forming the rock. If a
smaller sample in comparison with grain size is used, individual
grains will play too important a role in determining the properties
of the sample, and the results will not be representative of the
rock in bulk. The accuracy of a measurement may be seriously
affected by surface current conduction. Therefore, without
consideration of which measurement system is used, guard rings
should be employed which eliminate the flow of measured current
over the outside of the sample and which also assure a more uniform electric field in the sample. The guard rings should be
placed no more than 2 mm from the measurement electrodes or
on the side surface of the sample. In the latter case, the guard
rings may be steel or copper wire wrapped around the sample in
several turns. For best contact, the wires should be coated with
a graphite emulsion.

In the two-electrode method, the electrodes are located
symmetrically with respect to each other on opposite sides of the
sample. One electrode, the high-voltage electrode is connected to
the current supply, while the other is connected to the measuring
meter. In using an electrode arrangement which includes a guard
ring as shown in Fig. 20, the electrodes should have one of the
following diameters: measuring electrode – 10, 25, or 50 mm;
high-voltage electrode – 20, 40, or 70 mm, respectively.

If measurements are to be made on a low resistance sample,
such as an ore mineral, it is preferable to make use of an ammeter connection as shown in Fig. 21a or b. The first circuit is
preferable when the output resistance of the voltmeter is large
in comparison with the sample resistance, while the second circuit is preferable when the ammeter resistance is small in comparison with the sample resistance. Also, various types of bridge
circuits may be used in direct-current determinations of low resistances.

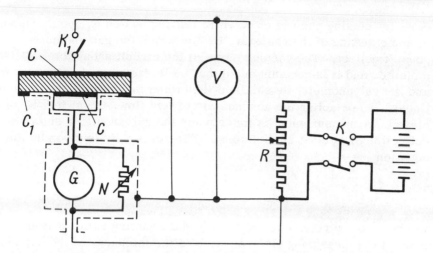

Fig. 20. Circuit for measuring the volume resistivity of a dielectric using a galvanometer: C — electrodes, C_1 — guard ring.

When measurements are being made on highly resistant
samples, a high-sensitivity galvanometer may be used in place of
the ammeter. The essential features of such a measuring circuit
are shown in Fig. 20. The circuit includes a voltmeter, V, for
measuring the voltage applied to a sample, a limiting resistance,
R, a galvanometer, G, with shunt, N, which may be varied over a
10,000 to 1 shunting range, and the rock sample under study. The
resistance of the shunted galvanometer must be adjusted to obtain
the proper speed of deflection. Without the shunting resistance,
the measurement is difficult.

A measurement is made in the following manner. First of
all, the current supply must be connected to the sample. This is

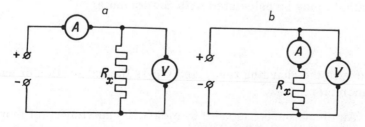

Fig. 21. Circuit for measuring resistivity using an ammeter and a
voltmeter.

done by closing the switch K while leaving switch K_1 open. There is some setting of the rheostat, R, for which the galvanometer doesn't deflect. The various parts of the circuit should be well isolated, and it is recommended that the measuring electrodes and the galvanometer be shielded (shielding is shown by the dashed lines). To establish the absence of current flow, the switch K_1 is closed. A measurement is made when the galvanometer shows a deflection of at least 20 divisions. The resistivity is given by the equation

$$\rho = \frac{VSn}{\alpha C_D h} ,$$

where V is the voltage in volts, n is the shunting ratio, S is the area of the measuring electrodes, h is the thickness of the sample, C_D is the calibration for the galvanometer, and α is the galvanometer deflection. The calibration factor, C_D, must be determined at the beginning of a measurement. Procedures for determining C_D may be found in an earlier monograph [79].

A series meter type E6-M may be used in measuring resistances in the range from 10^3 to 10^{12} Ω with direct current with the following accuracy: in the range 10^4 to 10^9 Ω, the error is no more than ±1.5%; to 10^{10} Ω, ±2.5%; to 10^{11} Ω, ±10%; and for 10^{12} to 10^{13} Ω, the accuracy is poorer. This meter permits the measurement not only of the initial and terminal resistances, but also the true resistances, using the method of averaging resistances measured with reversed current [67].

In measuring resistances of the order of 10^{15} to 10^{16} Ω, the current flowing through the sample for a length of time of the order of 300 sec is used to charge a capacitor. This charge is then measured with a ballistic galvanometer. The resistance of the sample may be calculated with the formula

$$R_x = \frac{V\tau n}{C_B \alpha} ,$$

where τ is the charging time, and C_B is the calibration of ballistic galvanometer.

An electrometer is used to measure resistances of more than 10^{16} Ω. The essentials of such a measurement method are shown in Fig. 22. For a measurement, the total capacity, C, of

the circuit is determined and multiplied by the reading of the electrometer in volts. At the time of opening of the switch K, the electrometer is charged by the charge flowing through the sample. If the
voltage V, applied to the sample and the deflection α of the electrometer in volts after a time τ following the opening of switch K are
known, the resistance of the sample may be computed with the
formula

$$R_x = \frac{V}{I} = \frac{V\tau}{C\alpha}.$$

A more complete discussion of methods may be found in
OST 40132.

Four-Electrode Method. In this method, the voltage
drop between two points or equipotential surfaces which lie between
the current contacts is measured. With this method, the effects
of electrode polarization are avoided. Therefore, not being subject to high-voltage polarization, the method may be used to
measure the true resistance of a sample. A variety of sample
shapes may be used in the four-electrode method, including cylinders, parallelepipeds, and cubes as well as irregularly-shaped
samples. Necessary requirements for the application of the four-
electrode method, as well as the two-electrode method, are that
good contact be obtained between the electrodes and the sample,
and that the sample be reasonably large in comparison with grain
sizes. The measuring electrode separation must also be several
times larger than the maximum grain size.

Figure 23 shows two typical ways in which electrodes may
be attached to a sample in a four-electrode method. One pair of
electrodes, A and B, is used to provide current circuit. The
other pair of electrodes, M and N, is used to measure voltage.
The measuring electrodes are formed from wire held against the
surface of the sample, or are probes with point contacts. The
M – N electrodes usually are placed at a distance of more than
10 mm from the current electrodes. Knowing the amount of current flowing in the sample and measuring the voltage drop between
the measuring electrodes, the resistivity of the sample may be
calculated using the formula

$$\rho = \frac{\Delta V}{I} \frac{S}{l},$$

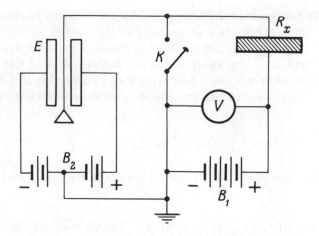

Fig. 22. Block diagram for a circuit measuring high resistance: R_X — sample being studied, B_1, B_2 — batteries, V — voltmeter.

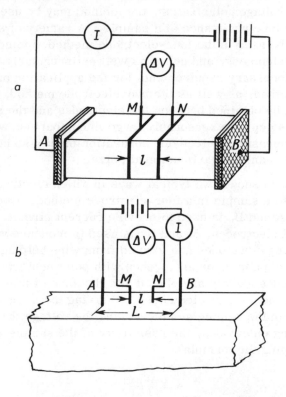

Fig. 23. Four-electrode arrays for measuring resistivity.

where S is the cross sectional area of the sample, and l is the spacing between the measuring electrodes (in scheme a).

The second scheme (b) for arranging a four electrode system in a linear array is often used for rapid determinations in the field. With this scheme, the resistivity is calculated with the formula

$$\rho = \frac{\Delta V}{I} \frac{\pi}{4} \frac{L^2 - l^2}{l} ,$$

where L is the separation between electrodes A and B, and l is the separation between electrodes M and N.

If four-electrode measurements are being made on high-resistivity samples, surface conduction may cause serious errors, and it may be necessary to use a guard ring between the measuring electrodes.

Comparison of resistivity values measured on the same samples with two- and four-electrode methods using direct current as well as with an alternating current method indicates that there is some scatter to the values determined with the various methods (Table 9).*

The data in Table 9 indicate that the values obtained with a four-electrode method fall between those obtained with the two-electrode method and with the AC method, or are close to the values measured with the AC method. This is caused by the fact that the variations between the values are due to effects at the electrode contacts which tend to increase the resistivity determined with the two-electrode method. With the AC method, electrode effects are minimized, and so the measured values of ρ are lower than those obtained with the two-electrode method. The matter of a possible variation of resistivity with frequency will be discussed in a later chapter (Chapter IV).

In addition to the basic methods for measuring resistivity which have been described here, there are numberous other methods not so commonly used [83-85].

Electrical Resistivity of Minerals

In considering the physical properties of some multicomponent aggregate, it is necessary to know the physical properties of each component, inasmuch as their properties determine the prop-

*Data obtained by K. A. Valeev and the author.

Table 9. Resistivity Values Determined with Different Techniques

Rock	Water content, % by wt.	Resistivity, 10^{-5} Ω-cm		
		with direct current		AC method f = 1000 cps
		two-electrode method	four-electrode method	
Dolomite	2.0	7.4	5.3	5.3
	1.3	6.0	6.0	5.5
	0.96	8.5	8.0	7.0
Graywacke sandstone	1.16	7.2	4.7	4.25
	0.45	110.0	58.0	58.0
Arkosic sandstone	1.26	2.22	1.05	0.96
	1.0	2.35	1.42	1.4
Organic limestone	11.0	0.87	0.6	0.4

erties of the aggregate. The wide range of values for the resistivity of rocks, which covers about 20 orders of magnitude, depends primarily on the electrical properties of the various minerals comprising rocks and the relative abundances of these minerals. However, water, when it is present in a rock, exerts a profound influence on the electrical properties, particularly in sedimentary rocks.

Table 10. Resistivity of Naturally Occurring Elements

Mineral	Resistivity, Ω-cm	Reference
Bismuth	12.7—14.3	[86]
Gold	0.22	[87]
Copper	0.17	[87]
Tin	1.31	[86]
Platinum	1.1	[87]
Mercury	9.6	[88]
Silver	0.16	[87]

All minerals may be classified as conductors, semiconductors, or dielectrics, depending on their conductivities. Most
minerals, as mentioned earlier, are crystalline and consequently
exhibit anisotropy. Anisotropy in the structure of a rock may be
explained by a variation in the physical properties and, in particular, in the electrical conductivity along the different crystallographic axes. The conductivity may be described with a symmetric tensor. This means that as a set of coordinate axes x, y, and
z are chosen along the principal crystallographic axes, the resistivity of the crystal may be expressed in terms of three values,
ρ_{11}, ρ_{22}, ρ_{33}. For a one-dimensional crystal, $\rho_{11} = \rho_{22} \neq \rho_{33}$.

If the crystal is oriented in some other direction with respect
to the coordinate axes x', y' and z' the conductivity tensor has the
form

$$
\begin{array}{cccc}
 & V_1' & V_2' & V_3' \\
I_1' & 1/\rho_{11}' & 1/\rho_{12}' & 1/\rho_{13}' \\
I_2' & 1/\rho_{21}' & 1/\rho_{22}' & 1/\rho_{23}' \\
I_3' & 1/\rho_{31}' & 1/\rho_{32}' & 1/\rho_{33}'
\end{array}
$$

Inasmuch as $\rho_{21}' = \rho_{12}'$, $\rho_{31}' = \rho_{13}'$, and $\rho_{23}' = \rho_{32}'$, the resistivity of a crystal may be defined by a set of six values. Knowing
the three resistivity components for the principal directions and
the cosines of the angles c_{ik} between the principal and any other
coordinate set, values for ρ_{ik}' may be determined from the formula

$$
\rho_{ik}' = \sum_{e=1}^{3} \sum_{m=1}^{3} c_{ei} c_{mk} \rho_{em}.
$$

Resistivity values which are listed in Tables 10-12 were
measured without respect to crystallographic directions, with few
exceptions. This may be one of the main reasons that values reported for a single mineral by different investigators vary.

Usually, in geophysical well logging, resistivity values are
expressed in ohm-meters, while in theoretical and laboratory
studies they are expressed in ohm-centimeters (1Ω-m is used as
a unit).

Table 11. Resistivity of Ore Minerals*

Mineral	Chemical formula	Resistivity, Ω-cm	Reference
Argentite	Ag_2S	$2 \cdot 10^{-1}$—10^6	[91]
Arsenopyrite	FeAsS	$1.10 \cdot 10^{-2}$—$6.0 \cdot 10^{-2}$	[92]
		$7.5 \cdot 10^{-2}$—$4.5 \cdot 10^{-1}$	[91]
Bornite	Cu_5FeS_4	$2.5 \cdot 10^{-3}$—$6 \cdot 10^{-1}$	[91]
		$5 \cdot 10^{-1}$—$6 \cdot 10^{-2}$	[22]
Braunite	Mn_2O_3	$1.6 \cdot 10^1$—$1.2 \cdot 10^2$	[91]
Bauxite	$Al_2O_3 n H_2O$	$2 \cdot 10^4$—$6 \cdot 10^5$	[91]
Bismuthinite	Bi_2S_3	$5.7 \cdot 10^4$	[91]
		$1.8 \cdot 10^3$	[22]
Wurtzite	ZnS	$3.5 \cdot 10^2$	[22]
Galena	PbS	$3.7 \cdot 10^{-2}$	[22]
		$2.6 \cdot 10^{-3}$	[22]
		8.7—17.5	[93]
		$6.8 \cdot 10^{-3}$—$3.2 \cdot 10^1$	[91]
Hematite	Fe_2O_3	0.35—0.7	[22]
		$2 \cdot 10^2$	[22]
		$1.2 \cdot 10^1$—$5.7 \cdot 10^5$	[91]
Graphite	C	0.1—10^3	[92]
Ilmenite	$FeTiO_3$	$2.2 \cdot 10^2$	[22]
Covellite	CuS	$1.2 \cdot 10^{-4}$—$5 \cdot 10^{-4}$	
		$1 \cdot 10^{-4}$—$2.8 \cdot 10^{-3}$	[22]
Cobaltite	CoAsS	1—5	[91]
		$3.5 \cdot 10^{-2}$	[92]
Cassiterite	SnO_2	$4.5 \cdot 10^{-2}$—$4.4 \cdot 10^2$	[91]
Cuprite	Cu_2O	0.1	[91]
Marcasite	FeS_2	2.8	[22]
		1.4—10	[91]
Magnetite	Fe_2O_4Fe	$3.6 \cdot 10^{-2}$—$5.7 \cdot 10^5$	[22]
		$8 \cdot 10^{-3} - 5 \cdot 10^{-1}$	[91]
Molybdenite	MoS_2	$7.5 \cdot 10^2$—$1 \cdot 10^6$	[22]
		$5.0 \cdot 10$—$1.6 \cdot 10^4$	[91]
Manganite	MnO [OH]	1.25—$3 \cdot 10^1$	[22]
Nicolite	NiAs	$2.0 \cdot 10^{-4}$	[91]
Pyrite	FeS_2	$2.3 \cdot 10^{-2}$—$1.5 \cdot 10^2$	[91]
		$2.9 \cdot 10^{-3}$—9.2	[91]
Pyrrhotite	$Fe_{x-1}S_x$	$6.5 \cdot 10^{-4}$—$4.1 \cdot 10^{-2}$	[22]
		$8.4 \cdot 10^{-3}$—10^{-1}	[91]
Pyrolusite	MnO_2	$1.6 \cdot 10^2$	[22]
Siderite	$FeCO_3$	$7.1 \cdot 10^3$	[22]
Chalcocite	Cu_2S	4.2	[91]
		$2.3.10^3$	[22]
Chalcopyrite	$CuFeS_2$	10^{-2}—$7 \cdot 10^{-2}$	[92]
		$1.17 \cdot 10^{-3}$—4.8	[91]

*Different values listed for a single mineral indicate samples taken from different sources.

Native elements occur to form conductive minerals. Average values for the resistivity of such elements which occur most commonly are listed in Table 10.

The presence of impurities in metals leads to variations in resistivity [89-90], just as in the case with semiconductors and dielectrics, therefore, the actual value of resistivity for a particular sample may deviate widely from the value listed in the table. Such impurities have an especially large effect on the conductivity of a material when they are distributed uniformly through the material (as for example, when the impurity is in solid solution). The resistivity in such a case is higher than the resistivity of either component. Native elements frequently form solid solutions, which is one of the reasons their electrical properties vary widely. The role of mechanical mixing of impurities on the resistivity of a metal is no less important than the role of chemical mixing.

Electrical conduction in native metals is by free electrons, which explains their high conductivity.

Another group consists of minerals with somewhat higher resistivities, causing them to be classed as semiconductors. This class includes the sulfide minerals (with the exception of antimonite, cinnabar, and a few others), arsenides, graphite, and a few of the oxides. The values of resistivity for these minerals which have been reported by various authors are listed in Table 11.

The values listed in Table 11 indicate that the same mineral taken from different sources will not necessarily have the same properties. The range in values for electrical resistivity may be as large as two or three orders of magnitude. The range in values for resistivity which one may observe for a single mineral is well illustrated by the distribution curves (Fig. 24) which have been compiled by Semenov [91]. Considering the small amount of research which has been done on the mechanism of conduction in minerals, such wide ranges in values can be explained only in the most general terms using solid-state theory. It has been established that the most important factor for all classes of conductors is the presence of impurities.

The effect of chemical impurities on the electrical conductivity of semiconductors was covered earlier. It is sufficient to

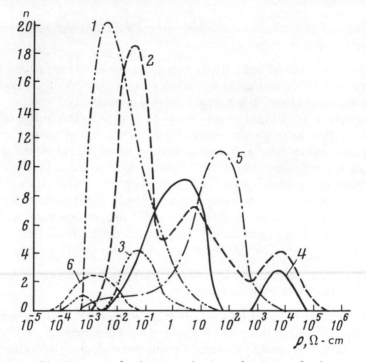

Fig. 24. Histograms for the reported values of resistivity for the
common minerals with high conductivity: (1) bornite, (2) magnetite,
(3) pyrrhotite, (4) arsenopyrite, (5) galena, (6) covellite; n — the
number of measurements.

reiterate that, independently of their chemical properties, impurities tend to reduce the resistivities of ionic dielectrics and electronic semiconductors, but tend to increase the resistivities of
metals. The degree to which an impurity will affect resistivity
depends on the quantity and on the physicochemical properties of
the impurity. Impurities mechanically mixed into the crystal
structure of a semiconductor also tend to reduce the resistivity.

The lowest values of resistivity within the semiconductor
group of minerals are shown by the following ore minerals: pyrite, galena, pyrrhotite, nicolite, chalcopyrite, and bornite. Intermediate to relatively high values of resistivity are reported for
wurtizite, ilmenite, tenorite, sphalerite, bauxite, magnetite,
wolframite, chromite, and others.

For most of the minerals listed in Table 11, only values of resistivity have been reported. The number of minerals for which electrical properties other than resistivity or the mechanisms of conduction have been studied is quite small. The most thorough studies are those which have been done on graphite [94-98], which has many uses in the electronics industry. Work has also been reported on the electrical properties of molybdenite [99-100], magnetite [101], wurtzite [102], pyrite [103, 104], galena, and chalcopyrite [93].

A large degree of anisotropy has been observed in the case of graphite. For current flow parallel to the basal cleavage, the reported values of resistivity are about 10^{-4} Ω-cm, a value representative of metallic conduction [94-96]. Various values for the resistivity measured normal to the cleavage have been reported. Dutta [95] and Krishnan and Ganguli [94] have obtained similar values in the neighborhood of 0.2 to 1.0 Ω-cm, . while Primak and Fuchs [96] have found resistivities two orders of magnitude smaller $(5 \cdot 10^{-2}$ Ω-cm). Such a range in values for resistivity might be explained by different contents of impurities. The properties of graphite have been studied not only in the monocrystalline form, but also in polycrystalline forms such as rods and pressed forms. The resistivity of polycrystalline graphite depends on crystal orientation, crystal size, and density. The density is particularly important in determining resistivity. As the density increases, the resistivity drops markedly [98]. The degree of anisotropy is smaller in polycrystalline graphite than in the monocrystalline form. According to Kunchin [98], the ratio of resistivities measured in two directions is about 1.7. In most cases, the Hall coefficient for graphite has a negative value [96, 97]. In studies of the Hall coefficient for samples of polycrystalline graphite as a a function of temperature for various grain sizes, it was found that above some critical temperature, the Hall coefficient approaches a limiting value which does not depend on the magnitude or sign of the Hall coefficient at lower temperatures. Such a behavior for the Hall coefficient may be explained, according to Kunchin, by supposing that conduction is by free electrons as well as by holes. According to Tyler and Wilson's results [97], the value for the Hall coefficient for polycrystalline graphite falls in the range 0.39 to 0.66.

Molybdenite, as well as graphite, exhibits a high degree of anisotropy. The values of resistivity measured in the two principal directions vary by three orders of magnitude [99]. On the basis of the variation of the resistivities measured in the two principal directions with temperature, Dutta suggests that there are two types of conduction in monocrystalline molybdenite – low-temperature and high-temperature. He considers that electrons contribute to conduction at the high temperatures. In another paper, a broad-scale investigation of the electrical properties of molybdenite has been reported, including measurements of Hall coefficient, conductivity, and thermoelectric coefficient. These properties were studied over the temperature range -183° to 500°C.

The observed sign for the Hall coefficient and the wide range of values observed with different samples indicate that molybdenite is a semiconductor whose properties are controlled by impurities. The sign of the thermoelectric coefficient indicates that the charge carriers are holes.

Dominicaly has studied the properties of natural and synthetic crystalls of magnetite [101]. It was established that the resistivities measured along the three principal crystallographic directions (111, 100, and 110) do not differ appreciably at room temperature, but are different at low temperature. The values for resistivity along the three principal directions at about -170°C are in the ratio $\rho_{111}: \rho_{100}: \rho_{110} = 5: 3: 2$.

The activation energy for synthetic wurtzite, determined from the temperature variation of resistivity, is 3.77 eV [102].

According to references [103] and [104], conduction in pyrite may be by holes or by electrons. The Hall coefficient determined at room temperature for n-type pyrite falls in the range 5.6 to 6.5, and for p-type pyrite, in the range 2.3 to 6.3. The activation energy for pyrite, determined in the region of intrinsic conductivity, is 1.2 eV [104].

Galena from Azerbaidzhan, according to [93], exhibits both impurity and intrinsic conduction over the temperature range 20-450°C, while chalcopyrite from the same area exhibits only impurity conduction over a broader range of temperatures , 20 to 650°C.

Information concerning the sign of the charge carriers in semiconducting minerals may be obtained by studying the thermo-electric properties as well as the Hall coefficient. Studies of thermoelectric properties are very useful in the solution of a number of geologic problems, as well as in explaining the nature of conduction [105-108]. They provide some information about the temperature of formation of some minerals for example.

Very extensive studies of the magnitude and sign of the thermoelectromotive force (thermo-emf) of a variety of ore minerals, including relatively rare minerals, have been reported by Telkes [109]. She found that in most cases several samples of a single mineral exhibit similar values of thermo-emf, but that a few minerals, such as galena and pyrite, show wide dispersion of measured values, not only in magnitude but also in sign. The largest negative thermoelectromotive forces (-400 to -1000 μV/ deg) were obtained with argentite, chalcopyrite, and some samples of galena and molybdenite. Large positive values of thermo-emf are exhibited by chalcocite, bornite, enargite, marcasite, stannite, and others. In addition, Noritomi [110] has measured not only thermo-emfs, but has also established the temperature dependence in the cases of three ore minerals – galena, pyrite, and magnetite. For all three, the thermo-emf had a negative sign, and the magnitude decreased in inverse proportion to the temperature. The negative sign indicates that conduction was by electrons over the temperature range studied, 0 to 300°C.

Practical application of thermo-emf data was first suggested by Smith [105]. He established an empirical correlation between thermoelectric potentials in pyrite crystals and the temperature of formation. Studies by Gorbatov [106] in this same area were even broader, inasmuch as they permitted correlation between thermo-emf and the content and nature of impurities in minerals, as well as with the peculiarities of crystal structure. It has been reported [107] that the magnitude of thermo-emf may be used to work out the sequence of ore mineralization. In this study, it was found that two samples of galena, representative of different stages of mineralization, had markedly different thermo-emfs.

Work by Fantsesson [108] in this area dealt with the thermoelectric properties of natural solid solutions in samples of ilmenite from the Yakut kimberlite. A correlation was found between thermo-emf and the quantity of trivalent iron ions in the basic ilmenite lattice, as well as a significant difference between the thermo-emfs observed for ilmenite samples taken from kimberlite and traprock. On the basis of this correlation, the author concludes that thermo-emf data are important in determining the particular composition of minerals which affect semiconductor properties.

Thus, a study of the various properties of minerals indicates that in this group semiconductors may have either electrons or holes as charge carriers. The impurities are most important; the type of impurity may lead to one type of conduction in one sample, to the other type of conduction in another sample.

The values for the electrical properties of ore minerals are of interest not only in the various electrical prospecting methods, providing geologic information, but also possibly for discovering semiconducting materials which may be of value in the electronics industry.

Minerals in the third group have very high resistivities, being classed as dielectrics. Many of them are widely distributed in the earth's crust in the form of rock-forming minerals. Data on electrical conductivities are listed in Table 12.

The electrical resistivity of minerals in this group vary over a wide range, depending on the source of the samples as well as on the quantity and chemical properties of impurities. We do not know of any detailed studies of the electrical conductivities of the minerals listed in the table with the exception of mica, whose properties in AC fields are discussed in a later section.

The dielectric minerals volumetrically are much more important than the metallic and semiconducting minerals, but the number for which electrical data are available, as listed in Table 12, is small. Moreover, in addition to interest in mechanisms of conduction in such minerals, the current interests in the electrical properties of the earth makes these minerals worthy of further study.

Table 12. Resistivity of Resistant Minerals

Mineral	Chemical formula	Resistivity, Ω-cm	Reference
Diamond	C	$5 \cdot 10^{14}$	[88]
Anhydrite	$CaSO_4$	10^{11}	[87]
Halite	NaCl	10^{12}—10^{15}	[24]
Quartz, $\perp$ to axis	SiO_2	$2 \cdot 10^{16}$	[88]
Quartz, $\parallel$ to axis	SiO_2	10^{14}	[88]
Calcite	$CaCO_3$	$5 \cdot 10^{14}$	[88]
Cinnabar	HgS	$2 \cdot 10^9$	[87]
Muscovite	$KAl_2[OH]_2 \cdot AlSi_3O_{10}$	10^{14}—10^{16}	[111]
Limonite	$Fe_2O_3 \cdot n H_2O$	10^9	[87]
Stibnite	Sb_2S_3	10^7—10^{14}	[22]
Sulfur	S	10^{17}	[88]
Sylvite	KCl	10^{13}—10^{15}	[24]
Phlogopite	$[Si_3AlO_{10}] \cdot [F, OH]_2$	10^{13}—10^{14}	[111]
Fluorite	CaF_2	$7.9 \cdot 10^{15}$	[87]

Effect of Mineral Composition on the Electrical Conductivity of Rocks

Studies of the mechanism of conduction in rocks and the effect of various parameters constitute a very large and difficult subject. At the present time, conclusions about the electrical conductivity of rocks have been drawn from a variety of investigations. These basically have established the dependence on mineral composition, liquid content and its electrical conductivity, and shape and size of mineral grains, texture, and so on. We will consider the role of each of these factors below. However, many of these correlations have been insufficiently studied.

The resistivities of rocks cover a very wide range, as has already been said. In a general sense, the resistivity of a rock is determined by the electrical properties of minerals, which, as we have seen, belong to one of three classes: conductors, semiconductors, and dielectrics. Depending on which type of mineral is more important in a rock, rocks may be divided into two groups.

One group consists of rocks containing large amounts of low-resistivity ore minerals and exhibiting resistivities in a range characteristic of semiconductors. This group of rocks includes ores. The other group consists of rocks made up mainly of minerals with high resistivity. The constituent which is most effective in reducing the resistivity of such rocks is the brine solution filling the pore structure. This, as well as the other group of rocks, can be considered as a first approximation to be two-component systems, the components being high-resistivity or low-resistivity. A model consisting of two components is the simplest for theoretical considerations.

There have been many attempts to calculate the electrical conductivity of aggregates consisting of grains distributed through a conductive or resistive matrix. The expressions so derived are commonly listed in textbooks on electrical exploration [86, 112]. In addition, basic formulas for computing the resistivity for a variety of two-component aggregates are developed and presented in a monograph by Kobranova [24].

Up to the present time, the models which have been studied theoretically include conductive grains in the form of spheres with uniform diameter, ellipsoids of revolution, three-dimensional ellipsoids, and cubes, etc., distributed in a resistive matrix. The opposite case, in which the disperse component is resistive, has been treated also. The first model is best suited to describing rocks with ore minerals, while the second is best suited to describing sedimentary and igneous rocks. Most of the known formulas may be expressed in the following form:

$$\rho_{1,2} = P_{1,2}\rho_1,$$

where ρ_1 is the resistivity of the distributed component, $\rho_{1,2}$ is the bulk resistivity of the rock, and $P_{1,2}$ is a parameter depending on the mineral composition and texture of the rock (a coefficient of proportionality between bulk resistivity and the resistivity of the distributed component).

The validity of the theoretical expressions has been verified experimentally in a number of cases for the appropriate models [113, 114]. One example of such a comprehensive study of the electrical resistivity of two-component aggregates is the work of Ovchennikov and Kilukova [114, 115]. They developed the follow-

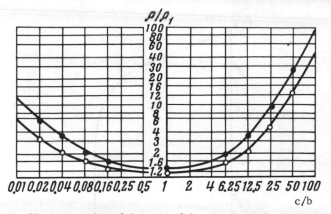

Fig. 25. Relationship of the ratio of the resistivity of pure shale, ρ, and shale with inclusions, ρ_1, to the ratio of semiaxes c/b for metallic inclusions: open circles — experimental data for a volume concentration of 0.05. Filled circles — the same for a volume concentration of 0.1.

ing formula for computing the effective electrical conductivity of a medium with the conductive component present as randomly distributed ellipsoids of a uniform size

$$\bar{\sigma} = \frac{\sigma_{xx} + \sigma_{yy} + \sigma_{zz}}{3},$$

where σ_{xx}, σ_{yy}, and σ_{zz} are the components for the tensor describing the effective electrical conductivity of the medium.

The validity of the theoretical formulas was checked by the authors using artificial materials. The distributed material was ground shale, and the dispersed component was elliptical sheet metal disks, pellets, and wire clippings of uniform size distributed randomly. The theoretical curves for the relation between the ratio of the resistivity of pure shale to the resistivity, ρ_1, with inclusions and the ratio of semiaxes, c/b, for two different concentrations of inclusions — 0.05 and 0.10 — are shown in Fig. 25. We see the excellent agreement between the experimental data and the theoretical curves, as well as the importance of the ratio of semiaxes, c/b, in concentration of the conductive components. This is in accord with theoretical and experimental results reported by Semenov [116, 117] concerning the effect of fabric on the resist-

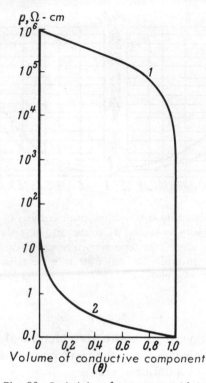

Fig. 26. Resistivity of aggregates with inclusions in the form of spheres: (1) ratio
of the resistivity of the inclusions to the
resistivity of the matrix, $10^{-1}/10^6$, (2)
ratio of resistivities, $10^6/10^{-1}$.

ivity of aggregates containing inclusions of isometric form and in
the form of ellipsoids of revolution. The following generalizations
may be drawn from the results of these investigations:

1. Above all, the resistivity of a rock depends on whether
the conductive component is present as the matrix or as the disperse phase. Two different typical characteristic relationships
between bulk resistivity and the volume fraction of conductive
mineral are found, depending on which type of rock is being
considered. Both cases are illustrated in Fig. 26. The upper
curve expresses the resistivity of a medium in which the conductive component is present in a disperse, discontinuous form, and
the lower curve expresses the resistivity of a medium in which the

conductive phase is present as the matrix. Thus, at low concentrations for a conductive material in an aggregate in which the matrix is resistive, a small increase in the amount of conductive material causes only a small decrease in bulk resistivity. However, in an aggregate in which the matrix is conductive, a small increase in the amount of conductive material evokes a large decrease in the bulk resistivity. The characters of the relationships for the two cases at high concentrations of the conductive phases are also different.

2. If the conductive inclusions are strongly elongate or needle form, the bulk resistivity will exhibit a well developed anisotropy. The lowest resistivity is observed along the long dimensions of the conductive grains. Also, the relationship between bulk resistivity and the volume content of conductive grains depends on direction. In one or two directions, depending on the shape of the inclusions, the relationship assumes the form characteristic of a medium with conductive inclusions; in other directions, on the other hand, the relationship assumes the form characteristic of a medium with resistive inclusions.

3. The resistivity of an aggregate with conductive cement depends weakly on the degree of homogeneity of the inclusions. In media with poorly conducting cement, the degree of homogeneity of the inclusions and their form are quite significant. The presence of small amounts of elongate conductive minerals along with conductive spherical inclusions may lead to a marked reduction in bulk resistivity.

From fundamental considerations about the electrical conductivity of two-component aggregates, it follows that the resistivity of an orebearing rock is a function not only of the amount of ore mineral in the rock, but also is very dependent on the fabric and interrelation between the ore minerals and the rock matrix. However, the number of studies in which electrical properties and fabric have been examined is not yet adequate. The most complete of such studies are those by Semenov [116, 117], Murashov et al. [118], and Parasnis [92].

These papers, especially the important study completed by Semenov [117], allow us to deduce the relationship between the resistivity of the more important economic ores and their mineral composition and fabric.

Pyrite Ores. On the basis of mineral composition, pyritic ores may be classified as sulfur pyrite ores, copper–sulfur pyrite ores, copper zinc, and copper–zinc arsenide ores. These ores commonly occur in the form of pods or veins with various shapes.

The mineral which determines the electrical conductivity of a pyritic copper ore is iron pyrite. The resistivity of such ores varies over a wide range depending on the amount of iron pyrite present and its structure. Iron pyrite usually crystallizes as small, distinct grains and affects the bulk resistivity of a rock only when it is present in relatively large amounts. If there is a large amount of gangue material (quartz, sericite), the pyrite grains may be electrically insulated from one another. In such a case, the grains are dispersed through the high resistivity gangue material and the resistivity of the ore is approximately that of the gangue material. Obviously, therefore, disseminated ores exhibit high resistivity and are not distinguishable from the host rock. With decreasing amounts of gangue materials, the films of insulating mineral between pyrite grains become thinner and some pyrite grains may touch, with the resistivity dropping. It should be emphasized that variations in the thickness of the insulating films between grains and variations in the contact area between pyrite grains lead to a wide variability in the resistivity of a given ore body.

If the dissemination of ore minerals has a boxwork or dendritic form, that is, if the pyrite grains are distributed in a dendritic network forming conductive paths, the resistivity of the ore is sharply reduced. In ores with a high content of noneconomic pyrite distributed in the interstices between the grains of the ore minerals, there would be no contact between the pyrite grains. In this case, the bulk resistivity would be close to that of the ore mineral, but obviously, somewhat higher.

The values of resistivity reported by Murashov et al. [118] for copper pyrite ores are listed in Table 13. From this table it follows that ores which contain 60% pyrite ore or more are markedly different in resistivity from ores with less pyrite. However, the high resistivity of the trans-Caucasus ore with a high pyrite content is an anomaly. The variation curve constructed by Bukhnikashvili [119] for the variation of the resistivity of copper pyrite

Table 13.   Resistivity of Copper Pyrite Ores

Region	Content, %					Resistivity, Ω-cm
	Pyrite	Chalco-pyrite	Sphalerite	Other ore minerals	Gangue	
Mariin Taiga	18	2			80	30,000
Trans-Caucasus	40			20	40	13,000
Mariin Taiga	60		5	15	20	90.0
Urals	75		10	5	10	14.0
Urals	95		5			10.0
Trans-Caucasus	95				5	700.0

from the trans-Caucasus measured in place also indicates a high resistivity, with ρ falling in the range $1.5 \cdot 10^3$ to $1.0 \cdot 10^4$ Ω-cm. Obviously, the explanation must lie in the distribution of pyrite grains, which are distributed ineffectually in a resistant matrix. According to data reported by Parasnis [92], the resistivity of copper-pyrite ores falls in the range from several tens of Ω-cm to two or three hundred Ω-cm for a pyrite content of 40 to 60%, while for larger amounts of pyrite, the resistivity decreases to 10^{-2} Ω-cm.

Pyrite Copper–Sulfur Ores are distinguished by somewhat higher conductivities than pyrite copper ores because of the presence of chalcopyrite, which has a resistivity of 10^{-2} to 5 Ω-cm. Filling the interstices between the quartz and pyrite grains, and in some cases, forming an emulsion in pyrite or other minerals, chalcopyrite contributes to a significant decrease in the bulk resistivity of the rock. According to Murashov et al. [118], pyritic copper–sulfur ores containing 60 to 90% chalcopyrite have resistivities between 10 and 70 Ω-cm, while data reported by Parasnis [92] indicate even lower resistivities, from 10^{-2} to 10^{-1}. It should be noted that increasing the content of chalcopyrite above 70% does not significantly reduce the resistivity. This is in accord with the theoretical relationship $\rho = f(\Theta)$ for a conductive mineral forming the matrix (the lower curve in Fig. 26).

Pyrrhotite Ores also are characterized by low values of re-sistivity. Pyrrhotite, as well as chalcopyrite, fills the interstices between grains and, therefore, forms conductive networks. The resistivity of pyrrhotite is 10^{-3} to 10^{-1} Ω-cm; that is, it is even lower than the resistivity of chalcopyrite. The resistivities of pyrrhotite ores with different pyrrhotite contents, as given by Parasnis are listed in Table 14.

These data indicate that pyrrhotite ores have the lowest re-sistivity of any of the ores which we have studied. However, we must keep in mind that these low values are reported from a sin-gle study in which resistivities reported for other ores are also exceptionally low. These low values may be a result of the meas-urement technique.

Essentially Zinc-Bearing Ores. The most important char-acteristic of ores of this type is the presence of a poorly conduc-ting ore minerals, sphalerite, which commonly occurs as a ce-ment about the other ore minerals. For example, sphalerite oc-curring with galena or pyrite frequently acts as a dielectric gangue material, like quartz, calcite, and other minerals. As a result, the resistivity in such rocks is controlled by the matrix resistivity. On the basis of several research studies carried out on a number of galena–sphalerite ores from the Caucasus, the resistivity of such ores lies in the range of 10^4 to 10^5 Ω-cm [119]. Ores in which the conductive mineral grains are insulated from one another by sphalerite and gangue minerals are not typical. Com-monly, other sulfides such as chalcopyrite, bornite, chalcocite, and others occur as continuous networks, markedly reducing the bulk resistivity. Moreover, chalcocite frequently occurs as an emulsion in sphalerite, lowering the resistivity of the mineral.

Considering the different characters of interrelations between minerals, that is, the fabric of the rock and various types of min-eralogic associations, the electrical conductivity of zinc ores varies over a wide range (five orders of magnitude or more). The amount of sphalerite is not an important factor in determining the bulk resistivity of the ore, which may be seen from the data listed in Table 15, as given by Murashov et al. [118].

Polymetallic Sulfide Suites are quite varied both in mineral composition and in fabric. The primary ore minerals contained in these suites are sphalerite and galena. In addition, polymetallic

Table 14. Electrical Resistivity of Pyrrhotite Ores

| Content, % | | Resistivity, 10^2 Ω-cm |
Pyrrhotite	Gangue minerals, quartz	
41	59.0	2.2
58	42.0	2.3
79	21.0	0.14
82	18.0	0.85
95	5.0	0.14

sulfide mineral suites may include pyrite, chalcopyrite, cassiter-
ite, boulangerite, molybdenite, silver sulfides, and others. The
presence of such poorly conducting minerals as molybdenite, boulangerite, and tetrahedrite in such ores, as well as the formation
of anglesite and cerrusite along grain boundaries decreases the
bulk conductivity of the ore. On the other hand, the presence of
chalcopyrite, pyrrhotite and arsenopyrite as coatings on the sphalerite significantly increases the bulk conductivity of the ore. The
gangue minerals may be quartz, calcite, barite, dolomite, or
others.

Various types of interrelations between the ore minerals and
the gangue minerals lead to a variety of fabrics, among which the
most important types are granular, banded, relic, coating of
sphalerite by galena, intergrown, and disseminated [117]. The
variety of fabrics for polymetallic sulfides ores causes a wide
range in values for electrical conductivity. Ores from the Caucasus, which contain pyrite, chalcopyrite, galena, sphalerite, and
gangue, have resistivities ranging from 11 to $5 \cdot 10^7$ Ω-cm [119].
In other ore deposits with polymetallic sulfide suites, resistivities
cover an even wider range from 10^{-3} to 10^8 Ω-cm [117].

Iron Ores. The most important mineral in iron is magnetite. In basic rocks, the magnetite may be associated with il-
mentite, while hematite may be present in some ores. In hydro-
thermal iron ores, the dominant mineral is siderite, along with
magnetite and hematite. In cases in which magnetite is more a-
bundant, the resistivity may be quite small, while in cases in
which siderite is more abundant, the resistivity may be high.

Table 15. Resistivity of Zinc Ores

Region	Sphalerite	Chalcopy-rite	Galena	Pyrite	Gangue	Resistivity, Ω-cm
Armenian SSR	30		5	15	50	$7.5 \cdot 10$
Trans-Caucasus	70	3	17	10		$2.0 \cdot 10^3$
"	80		10	10		$1.7 \cdot 10^5$
Armenian SSR	80	2	1	2	15	$1.3 \cdot 10^2$
"	90		5		5	$1.3 \cdot 10^4$

The electrical conductivity of iron ores depends on fabric to a large extent [117]. When the magnetite serves as the matrix around the non-ore minerals such as plagioclase, pyroxene, or others, the ore has a low resistivity. However, if high resistivity films between the grains develop, the resistivity of the ore will be much higher. Some contact metamorphic ores are excellent conductors, consisting of magnetite grains with connecting sulfide filaments, as well as ores consisting of serpentine, skarn, and magnetite.

Igneous titanomagnetite ores, in which magnetite is dispersed in diabase, are characterized by relatively large resistivities, as are fine-grained magnetite ores in which there are no sulfide minerals to form conductive filaments connecting the magnetite grains. Sedimentary iron ores and most metamorphic ores have a high resistivity. The iron minerals in sedimentary ores are such nonconductors as red and brown iron oxides, siderite, and iron chloride, while the iron minerals in metamorphic ores are martite, magnetite, and hematite. The resistivities for a number of iron ores are listed in Table 16.

Copper and Nickel Sulfide Ores are examples of ores in which a small amount of ore mineral (3 to 10%) renders a rock highly conductive. For the majority of copper-nickel ores, the resistivity is no more than 10^{-2} to 10^{-3} Ω-cm [117]. The uniformity of resistivity and its low value may be explained as follows. First, most of the minerals found in these ores (pyrrhotite, chalcopyrite, niccolite, and others) have very low resistivities; secondly, the minerals are distributed between the silicate grains, forming continuous box-works and vein structures. However, even considering the excellent conductivity of copper-nickel ores, it may be difficult to distinguish them from graphitic slates in electrical exploration. Graphite, as well as chalcopyrite, bornite, and covellite, tends to form continuous conductive filiments, and so markedly reduces the bulk resistivity of a rock [117].

Considering the data which have been examined for the most economically important and abundant ore minerals, it is apparent that textural factors are quite important, as well as mineral composition, and, moreover, studies of the resistivity in conjunction with studies of chemical and minerals composition and texture and genesis of rocks would be fruitful and useful in support of electrical exploration. Also, it should be realized that moisture

Table 16. Resistivities of Iron Ores

Region of deposit	Composition	Resistivity, Ω-cm	Reference
Altai	60% magnetite, with secondary iron minerals	$4.5 \cdot 10^2$	[118]
Caucasus	Magnetite ore from a contact metamorphic deposit	$5 \cdot 10^1 - 10^4$	[117]
Vakidshvari	Magnetite	$1.6 \cdot 10^2 - 2.2 \cdot 10^2$	[120]
Altai	Disseminated brown iron oxide	$8.0 \cdot 10^4 - 3.0 \cdot 10^8$	[118]
"	Brown oxide, 75% } Gangue, 25% }	$2.0 \cdot 10^6 - 8.0 \cdot 10^7$	[118]
Georgian SSR	Continuous fine-grained hematite	$2.5 \cdot 10^5$	[119]
Vakidshvari	Magnetite	$5.5 \cdot 10^5 - 8.5 \cdot 10^5$	[120]
Georgian SSR	Magnetite in pegmatite	$7.0 \cdot 10^5 - 2.0 \cdot 10^7$	[119]

content is an important factor in ores which have a resistivity higher than 10^4 Ω-cm. Therefore, the fact that the resistivity of such ores is often higher when measured in the laboratory than when measured in the field may be explained by the effect of moisture [118-122].

Experimental studies of the electrical properties of ores as a function of the content, textures, and size of conductive minerals are subject to some difficulties. As a result, it is necessary to resort not only to theoretical studies, but also to modeling. As earlier, theoretical formulas may be verified using models. Moreover, model studies permit the independent evaluation of the effect of such parameters as grain shape, size, and ore content on the resistivity of synthetic ore. Such synthetic rocks are most satisfactory, because they allow quantitative control of the ratio of conductive to nonconductive components, construction of required textures, and, in general, regulation of the physical and chemical properties of the model over the desired ranges.

Mandel et al. [123] have conducted investigations of synthetic rock samples made up of lead particles, quartz sand grains, cement, and a saturating solution with various concentrations of NaCl. In these studies, an anomalous increase in bulk conductivity was observed as the lead content was increased from 2 or 3 to 8%. The existence of this anomalous behavior was further substantiated in a more recent paper [124] in which the induced polarization effect was observed. The use of model studies not only offers novel possibilities for studying well-known phenomena and relations, but also makes explanation of these effects much simpler.

The probable limits for electrical resistivity in various types of ores which have not been listed in the preceding tables are given in Table 17.

Comparison of the ranges of variation in resistivity for ores and for ore minerals indicates that bulk ores always have higher resistivities than the minerals comprising the ores. The presence of a nonconducting component in an ore, which in many cases separates the grains of ore minerals, increases the ore resistivity in comparison with the resistivity of the pure ore mineral. Data reported by Parasnis [92] which indicate the same relation between mineral resistivity and bulk ore resistivity are listed in Table 18.

Rock with High Resistivity. High resistivity rocks include igneous rocks, metamorphic rocks, and a few sedimentary rocks, composed mainly of insulating minerals with insignificant amounts of moisture. Because of the low mineral conductivity and the low moisture content, the resistivities for such rocks are much higher than the resistivities for ores or for rocks with large water contents.

The electrical resistivity of these rocks is virtually independent of mineral composition and genesis, which may be explained as follows. Most of the rock-forming minerals have about the same resistivity. Thus, sedimentary rocks commonly have nearly the same resistivity as gneiss, quartzite, diabase, and basalt. The presence of only small amounts of moisture greatly decreases the resistivity of the most resistant rocks, and this in itself reduces the dependency of resistivity on mineral composition (see Table 19). If moisture is completely removed, there is a tendency for the resistivity of "absolutely dry" rocks to decrease in going from acidic to basic or ultrabasic composition. This is explained by the fact that the minerals olivene and pyroxene have somewhat lower resistivities than quartz and feldspar. Moreover,

Table 17. Resistivities for Various Ores

Rock	Region of deposit	Mineral composition, %	Resistivity, Ω-cm	Reference
Antimonite ore in quartz	USSR, Georgian SSR		$3.8 \cdot 10.5 - 3.5 \cdot 10^9$	[119]
Arsenopyrite ore	USSR, Chelyabin area	Arsenopyrite, 60; pyrite, 20; quartz, 20	$3.9 \cdot 10^1$	[118]
Arsenopyrite ore	Sweden		$1.1 \cdot 10^{-2} - 1.43$	[92]
Bornite ore	Australia	Rich	0.3	[22]
Bornite ore	USSR, Uspen area	Bornite 40, quartz 60	7.0	[118]
Wolframite ore	USSR, Urals	Wolframite, 80	$1.6 \cdot 10^6 - 1.9 \cdot 10^6$	[118]
Mineral complex with wolframite and cobaltite	USSR, Caucasus		$10^5 - 10^9$	[119]
Galena ore	USSR, Central Kazakhstan	Massive	7.0	[118]
" "	Sweden	Galena, 50–80	$1 - 3 \cdot 10^2$	[92]
" "	USSR, Caucasus	Nearly massive galena	82.0	[119]
Hematite ore	Sweden		$10 - 3 \cdot 10^4$	[92]
" "	USSR, Georgian SSR	Massive, fine-grained	$2.5 \cdot 10^5$	[119]
Graphitic slate	Norway		13.3	[22]
Massive graphite			$10^{-2} - 0.5$	
Molybenite ore	USSR, Armenian SSR		$1.6 \cdot 10^4 - 4.0 \cdot 10^5$	[119]
Pyrolusite ore	USSR, Nikolpol'	colloidal	$1.61 \cdot 10^2$	[118]
Chalcocite ore	USA, Jerome, Ariz		3.1	[22]
Chalcopyrite ore	Sweden		0.01–10	[92]

Chalcopyrite ore	USSR, Trans-Caucasus	Chalcopyrite, 80; pyrite, 10; gangue, 10	$6.6 \cdot 10^1$	[118]
"	Same	Chalcopyrite, 90; rare pyrite, 2; quartz, 8	$6.5 \cdot 10^1$	[118]
Chromite ore	Sweden	Chromite, 95; serpentine, 5	10^5	[22]
"	USSR, Urals		$1.2 \cdot 10^6$	[118]

Table 18.   Resistivities of Ores and
Ore Minerals

Basic mineral	Resistivity, Ω-cm	
	Ore	Pure mineral
Pyrite	0.01—1000	0.005—5
Chalcopyrite	0.01—10	0.01—0.07
Pyrrhotite	0.001—0.1	0.001—0.005
Arsenopyrite	0.1—10	0.03
Galena	1—30 000	0.003—0.03
Magnetite	1—1100	0.01

basic rocks contain a higher fraction of conductive heavy minerals
than acidic rocks.

Frequently, thin layers of magnetite are found in serpentinized peridotite, pyroxenite, and dunite, and also in gabbro and
basalt. These layers markedly reduced the bulk resistivities of
these rocks. If such ore minerals are absent from basic rocks,
they are highly resistant both in the wet and dry state. Introducing metallic minerals, graphite, or clay minerals into syenite,
gabbro, quartzite, andesite, or gneiss also reduces their resistivity. It is not possible to make a more detailed analysis of the
effect of mineral composition on resistivity in view of the lack of
adequate data. As may be seen from the data listed in Table 19,
in a number of cases there is a wide range of reported values for
a single mineral composition, with the range probably representing
variations in moisture content.

Moreover, these data probably include consistent errors in
measurement. The various types of measuring equipment probably
had degrees of precision, so that different sources may report
different resistivity values for the same rock. For measurements
made in the laboratory with direct current, it may not be reported
whether a resistivity value is the initial, stationary or true value.
Moreover, resistivity values measured with an alternating current
are commonly somewhat lower than the values measured with
direct current. Thus, it is necessary also to consider the relationship between resistivity and frequency. As a result, there are a

Table 19. Resistivities of Rocks

Rock type	Source	Resistivity wet	Resistivity dry	Reference	Comments
Sedimentary rocks					
Dolomite	Armenian SSR	$3.5 \cdot 10^4 - 5.0 \cdot 10^5$		[119]	Dolomitized form
Limestone	Kazakhstan SSR	$4.2 \cdot 10^7$	$1.2 \cdot 10^9$	[118]	
"	Georgian SSR	$2.1 \cdot 10^7$	$2.3 \cdot 10^9$	[119]	Fine-grained rocks
Limestone marl	Georgian SSR	$8.4 \cdot 10^8$		[119]	
Sandstone	Donbas	$1.41 \cdot 10^7$	$6.4 \cdot 10^{10}$	[125]	Water content, 0.37%
"	Kara-Shishak	$3.5 \cdot 10^6$	$3.1 \cdot 10^7$	[118]	
Arkosic sandstone	Jezkazgan	$6.8 \cdot 10^4$	$1.0 \cdot 10^8$	[118]	
Quartzitic sandstone	Dashkesan	$2.3 \cdot 10^4 - 3.3 \cdot 10^4$		[120]	
Tuffaceous sandstone	Dashkesan	$10^5 - 1.0 \cdot 10^7$		[120]	
Metamorphic rocks					
clay slate	Georgian SSR	$6.4 \cdot 10^6$	$1.6 \cdot 10^7$	[119]	strongly quartzitic
"	"	$1.1 \cdot 10^6$	$1.6 \cdot 10^9$	[119]	
"	Nerchinsk	$1.0 \cdot 10^5$	$1.0 \cdot 10^8$	[118]	
"	Georgian SSR	$4.0 \cdot 10^5$	$6.0 \cdot 10^8$	[119]	shaly

Rock type	Source	Resistivity		Reference	Comments
		wet	dry		
Quartz–sericite slate	Ziryanskii mine	$5.0 \cdot 10^6$	$3.6 \cdot 10^9$	[118]	
Quartz–chlorite slate		$5.0 \cdot 10^5$	$2.0 \cdot 10^8$	[118]	
Calcareous quartzite	Altai	$4.0 \cdot 10^5$	$2.0 \cdot 10^{10}$	[118]	
Quartzite	Georgian SSR	$4.7 \cdot 10^8$		[119]	
Metamorphosed tuff	Altai	$2.0 \cdot 10^5$	$1.0 \cdot 10^7$	[118]	
Gneiss	Kugrasin	$6.8 \cdot 10^6$	$3.2 \cdot 10^8$	[118]	
Marble	Nerchinsk	$1.4 \cdot 10^6$	$2.5 \cdot 10^{10}$	[118]	
"		$7.06 \cdot 10^{11}$	$1.8 \cdot 10^{20}$	[126]	
Hornfels	Kugrasin	$6.0 \cdot 10^7$	$6.0 \cdot 10^8$	[118]	
"	Trans-Caucasus	$8.1 \cdot 10^5$	$6.0 \cdot 10^9$	[118]	
Skarn	Armenian SSR	$2.5 \cdot 10^4$		[119]	
Igneous rocks, Acidic					
Granite	Azerbaidjan SSR	$3.0 \cdot 10^7$	$3.2 \cdot 10^{18}$	[119]	
"	Ubinskoe	$0.36 \cdot 10^9$	$0.3 \cdot 10^{16}$	[126]	
"	Kola Peninsula	$0.16 \cdot 10^9$	$1.3 \cdot 10^8$	[126]	
Granite porphyry	Urals	$4.5 \cdot 10^5$	$1.0 \cdot 10^7$	[118]	
Granite porphyry		$7.0 \cdot 10^5$		[118]	
Quartz vein	Komsomol'skoe	$0.6 \cdot 10^{10}$	$1.0 \cdot 10^{18}$	[126]	
Quartz porphyry	Georgian SSR	$9.2 \cdot 10^7$		[119]	
Feldspar porphyry	Australia	$4 \cdot 10^5$		[22]	
Leucophyre (albite)	Georgian SSR	$2.9 \cdot 10^4$	$0.4 \cdot 10^9$	[127]	
"	Urals	$4.5 \cdot 10^5$		[117]	

Rock	Locality			Ref.
Intermediate				
Porphyrite	Georgian SSR	10^3	$3.3 \cdot 10^5$	[119]
"	Azerbaidjan SSR	$6.7 \cdot 10^4$		[119]
"	Dashkesan	$5 \cdot 10^6$		[120]
Diorite porphyry	Caucasus	$1.9 \cdot 10^5$	$2.8 \cdot 10^6$	[118]
Carbonitized porphyry	Armenian SSR	$2.5 \cdot 10^5$	$5.9 \cdot 10^6$	[119]
Diorite	Urals	$2.8 \cdot 10^6$		[117]
Quartz diorite	Azerbaidjan SSR	$2.0 \cdot 10^6$	$1.8 \cdot 10^7$	[119]
"	Georgian SSR	$2.0 \cdot 10^8$		[120]
Dacite	"	$2.1 \cdot 10^6$		[119]
Andesite	"	$4.5 \cdot 10^6$	$1.7 \cdot 10^4$	[119]
Basic and ultrabasic				
Diabase porphyry	Trans-Caucasus	$9.6 \cdot 10^4$	$1.7 \cdot 10^7$	[118]
Diabase	Karelaia	$3.0 \cdot 10^6$	$2.2 \cdot 10^{11}$	[118]
"	Georgian SSR	$3.8 \cdot 10^4$	$3.3 \cdot 10^7$	[119]
"	"	$2.9 \cdot 10^4$	$0.8 \cdot 10^9$	[127]
"	Sibaevskoe	$4.57 \cdot 10^9$		[125]
"	Blyavinskoe	$1.18 \cdot 10^7$	$1.0 \cdot 10^{12}$	[126]
"	Khibini	$0.16 \cdot 10^8$	$0.17 \cdot 10^{13}$	[126]
Olivine norite		$3 \cdot 10^6 - 6 \cdot 10^6$		
"		$10^5 - 6 \cdot 10^5$		
Basalt	Armenian SSR	$1.6 \cdot 10^5$	$1.3 \cdot 10^9$	[119]
"		$2.3 \cdot 10^6$		[125]
Peridotite	Berestovetskoe	$3.0 \cdot 10^5$	$6.5 \cdot 10^5$	[125]

number of details concerning measurement technique which may
affect the value of resistivity for a rock sample.

Most of the data listed in this table were obtained from laboratory measurements. These values of resistivity are in agreement with values measured at outcrops and in mine workings provided the moisture content is the same. For example, Tarkhov
(see Dortman [8]) has obtained excellent agreement for the resistivities of porphyries and quartz-sericite slate measured both in
the laboratory and in mine workings. The resistivity of samples
with water adsorbed from the atmosphere will not be the same as
the resistivity of the same rock saturated with natural formation
water, which is usually highly saline. The difference in resistivity may be several orders of magnitude. The resistivity of a
rock in the natural state depends not only on its composition but
also on the manner of distribution of ground water in the rock, the
salinity of the ground water, the degree of weathering of the rock,
and even on climatic conditions [8]. Igneous and metamorphic
rocks which lie above the water table commonly have a higher resistivity than the same rock at greater depths, where they are completely saturated with water. The resistivity of rocks, heavily
fractured by tectonism, is always less than the resistivity of the
same rocks where they have not been altered. For example, in
the Altai, the resistivity of porphyry, quartzite, and limestone
has been found to be $6 \cdot 10^3$ to $8 \cdot 10^4$ Ω-cm in a fractured zone, and
$(1.2$ to $2.5) \cdot 10^5$ Ω-cm in a nonfractured zone. Variations in resistivity as a result of tectonic processes are very clearly shown
by the data in Fig. 27. The right part of the profile corresponds
to a faulted area. A decrease in the resistivity of igneous and
metamorphic rocks is found also as a result of weathering. The
greatest decrease is observed in granite, gneiss, and carbonitized
slate. The resistivity of basic and ultrabasic rocks is somewhat
higher because there are fewer joints and openings in such rocks
to provide access for weathering agents.

In conclusion, it should be noted that the relation between
mineral resistivity and the bulk resistivity of highly resistant rocks
is different from that noted for ores. The resistivity of igneous
and sedimentary rocks is always less than that of the minerals
composing these rocks. This is a result of the presence of moisture in water-bearing rocks, while for dry rocks this is the result

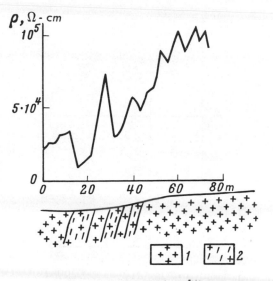

Fig. 27. Symmetrical electrical profiling curve over unaltered and serpentinized quartzite in a fractured zone of the Altai (from data by A. V. Veshev): (1) quartzite porphyry, (2) altered quartzite porphyry in fractured zones.

of the presence of significant amounts of impurities in the rocks, either mixed in or in chemical combination; it is well known that impurities reduce the resistivity of insulating materials.

Resistivity of Coal. The studies of the electrical properties of coals are of considerable importance, both theoretically and practically. The industrial utilization of electrical quality coke, the requirements of the electronic industry, and the use of geophysical exploration methods in the search for economic coal deposits all require a detailed knowledge of the electrical properties of coals. The data concerning the resistivity of coals which may be found in the literature do not always indicate the moisture conditions under which measurements were made, and many of the data were not obtained with adequate precision of measurements. These factors lead to a scattering of data which makes interpretation difficult. Data reported by various authors for six types of coal are listed in Table 20.

Table 20. Resistivity of Coals in Ω–cm [128]

Coal quality	Source of data			
	A. P. Kovalev [129]	V. M. Dakhnov [130]	A. A. Agroskin and I. G. Petrenko [131]	N. N. Shumilovskii [132]
Lower Moscovian Brown	$0{,}9 \cdot 10^{12}$	$10^3 - 2 \cdot 10^4$		$10^8 - 10^{10}$
Gaseous I*	$1{,}7 \cdot 10^{13}$		$10^9 - 10^{10}$	$10^8 - 10^{10}$
PSh	$4{,}9 \cdot 10^{12}$	$10^4 - 10^6$	$2 \cdot 10^{10}$	
K	$0{,}8 \cdot 10^{12}$		10^{10}	
PS	$2{,}6 \cdot 10^{11}$	$10^3 - 10^4$		
Anthracite	$3{,}0 \cdot 10^1$	$10^{-1} - 10^3$	10^2	

*Coal designations are defined in the text.

In reference [129], high values of resistivity are reported for measurements made on samples obtained from a coal pile in a completely dry state. Lower values of resistivity are reported for naturally wet coal [130], while the resistivity of air-dried coal is of about the same order of magnitude according to two references [131, 132].

Thus, the range of resistivity for coals is extremely wide, wider than the ranges for ores or for highly resistant rocks, with the range extending from the semiconducting to the insulating regions. The resistivity is a function of the physical and petrographic character of the coal, the mineral composition and the degree of metamorphism [128, 133–138]. Generally, the resistivity of a coal is specified by a correlation with texture and composition, but is a function of a greater number of variables than are the properties of other rocks.

The basic petrographic constituents of coals are fusain, vitrain, durain, and clarain. The resistivity of dry samples of clairain and vitrain is more than $3 \cdot 10^8$ Ω-cm, while fusian has a somewhat lower resistivity, 10^3 to 10^5 Ω-cm. As a result, the resistivity of a coal will depend on the relative proportions of these constituents. For example, the resistivity of a sample of Karagandin coal composed solely of vitrain is 10^{10} Ω-cm, while another sample of the same coal composed solely of fusain has a resistivity of $5 \cdot 10^8$ Ω-cm. The resistivity of a sample with an intermediate composition is 10^9 Ω-cm.

The main factor in determining the resistivity of a coal is the degree of metamorphism, which causes differences in carbonitization and carbon ratio [128, 134, 138]. With progressive carbonitization, the physical and chemical properties of coal change from those of a brown coal to those of an anthracite. Coals have been widely classified according to the degree of carbonitization as follows:

1. Brown coal (B)
2. Long-burning coal (D)
3. Gassing coal (G)
4. Greasy steam coal (bitumenous) (PSh)
5. Coking coal (K)
6. Lumpy steam coal (super-bitumenous) (PS)

7. Lean coal (sub-anthracite) (T)

8. Anthracite (A)

In this classification, the carbon ratio increases and the ash content decreases in going from the top to bottom. The changes in character caused by this metamorphism have the following effects on the electrical properties. Weakly metamorphosed coals (brown coals) have a rather high resistivity (10^{10} to 10^{11} Ω-cm) when dry, and 10^5 to 10^6 Ω-cm when moist. With increasing metamorphism, the resistivity increases to a maximum. For example, a cannel coal may have a resistivity of the order of 10^{12} Ω-cm [128]. Further metamorphism leads to the formation of free carbon stringers and a reduction of the organic radicals in the coal, causing the resistivity to be lower. According to reference [128], the resistivity decreases markedly at carbon contents greater than 87%. Similar results have been obtained from measurements on a large number of anthracite samples from the Donets Basin. Transition from sub-anthracite to anthracite coal causes a decrease in resistivity from 10^8 Ω-cm to 3 to 5 Ω-cm.

Dakhnov [139] has analyzed these data statistically, finding that the highest resistivities are associated with coals of type G and PSh (see Fig. 28).

According to Toporets [134], type D coals have the highest resistivities. Moreover, the amount of decrease in resistivity with increasing metamorphism depends on the type of metamorphism. A pronounced decrease in resistivity is found in coal fields subjected to thermal metamorphism, particularly contact metamorphism [134]. The disparity in results with respect to sub-bitumenous coals obviously is related to differences in the moisture contents of the investigated coals. It is obvious that dried-coal samples with low moisture contents were studied in reference [134], while the samples studied by Dakhnov had a higher moisture content.

The role played by the carbon ratio in determining the resistivity of coals with various degrees of carbonitization is not uniform, as is shown in reference [136]. For coals of types D, G, and PSh, the character of the relationship is one of decreasing resistivity with increasing carbon ratio, while in the case of anthracite, on the other hand, increasing the carbon ratio increases

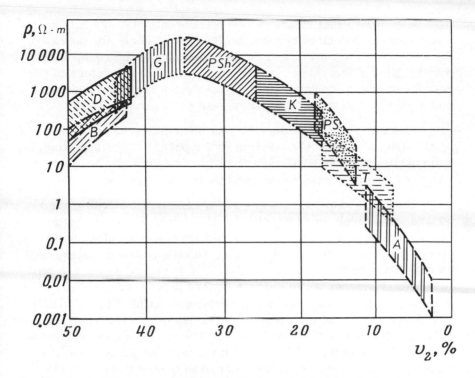

Fig. 28. Graph of the most probable values for the resistivity of coals of various grades as a function of quality. The ash content is plotted along the abscissa.

the resistivity. It must be noted that not only is the carbon ratio important, but so is the ash content, which is related to the carbon ratio. Kaolinite in a coal with an intermediate degree of carbonitization will markedly decrease the resistivity, while calcite or siderite will have little effect. In anthracite, the opposite situation is found.

Effect of Moisture Content on the Resistivity of Rocks

Water is a practically universal constituent of rocks. With the exception of a few ores, it strongly affects the resistivity of all rocks. The depressing influence of moisture on the resistivity of a rock can be explained by the fact that its resistivity is many orders of magnitude lower than the resistivity of most rock-forming minerals. The exact nature of the relationship between bulk resistivity and water content depends on the type of rock, the porosity, the salinity of the water, and the permeability of the rock. Rocks may be divided into three classes on the basis of the nature of the relationship between resistivity and moisture content; (1) dense and highly resistant, (2) clay-free, and (3) shaly.

Because the resistivity of water-bearing rocks is determined by the nature of the brine solution in the pores, its composition, and resistivity, it is necessary first to review what is known about various types of ground waters and their resistivities.

Resistivity of Natural Waters. Natural ground waters vary widely in composition. However, the most common constituent in all such solutions is sodium chloride, along with other salts and oxidates (carbonates, sulfates, phosphorates, and so on). Relatively low concentrations of dissolved salt are found in surficial waters, while connate waters may be very saline. The resistivity of a natural water is determined mainly by the salinity. The concentration of salt in solution varies over wide limits, from 0.1 mg/liter to 10 g/liter and more. The effect of chemical composition on the resistivity of an electrolyte is small because the mobilities for the ions commonly found in ground water are all nearly the same. Thus, it is possible to calculate the resistivity of a solution by assuming that all the salt in solution is the most common salt. Usually, we consider the relationship between resistivity and salinity for solutions of sodium chloride.

The resistivity of a solution may be calculated from the following formula:

$$\rho_v = \frac{10^3}{\Sigma(C_a l_a f_a + C_k, \, l_k, \, f_k)} \; (\Omega\text{-cm}),$$

where C_a and C_k are the concentrations, in gram equivalents, of the anions and cations in solution, l_a and l_k are the mobilities of the anions and cations, and f_a and f_k are the activity coefficients for the anions and cations, respectively [139].

The relationship between water resistivity and NaCl concentration at a temperature of 20°C is given in Table 21.

The salinites of natural waters and, therefore, their resistivities depend on rock composition and genesis, on climatic

Table 21. Resistivity of Solutions with
Various Concentrations of NaCl [8]

Concentration, g/liter	Resistivity, Ω-cm	Concentration	Resistivity, Ω-cm
0.005	$1.05 \cdot 10^5$	1.0	$5.8 \cdot 10^2$
0.05	$1.1 \cdot 10^4$	10.0	$6.5 \cdot 10$
0.5	$1.2 \cdot 10^3$	50.0	15

conditions, and on the relief of the terrain. In platform areas, the salinity varies in going from north to south from 0.1 to 0.5 g/liter (Baltic Shield) to 3 to 5 g/liter (Azov Massif) [8].

Lower resistivities are found for water in ore deposits and in tectonically active zones. In mountainous regions, ground waters have low salinities and high resistivities. Triple-distilled water has the highest resistivity, about 10^7 Ω-cm.

High -Resistivity Dense Rocks. The water content in igneous, metamorphic, and dense sedimentary rocks commonly is no more than 3%, and frequently is less than 1%. However, even this amount of moisture is enough that small variations in the amount will cause large variations in resistivity. Values for resistivity of various rocks as a function of water content, obtained by the author and from reference [125], are listed in Table 22.

It may be seen from these data that decreasing the moisture by a few tenths of a percent causes a change in ρ of an order of magnitude or more. Results of measurements by the author indicate that the slope of the line $\rho = f(w)$ is not the same for rocks of different genesis and petrographic composition (Fig. 29).

Resistivity increases most rapidly with decreasing water content in granites, and less rapidly in dolomite and basalt. It should be noted that experimental data points do not always lie along a straight line, but they rarely deviate very much from such a line. This reflects the fact that the rate of change of resistivity with water content varies not only with the nature but also with the size of grains, the structure of the pore volume, and the content of conductive minerals. In peridotite, for example, the low

Table 22. Resistivities of Rocks with Various Water Contents

Rock	Water content, %	Resistivity, Ω-cm	Rock	Water content, %	Resistivity, Ω-cm
Siltstone	0.54	$1.5 \cdot 10^6$	Diorite	0.02	$5.8 \cdot 10^7$
	0.5	$7.3 \cdot 10^7$		0	$6.0 \cdot 10^8$
	0.44	$8.4 \cdot 10^8$	Peridotite	0.03	$2.2 \cdot 10^6$
	0.38	$5.6 \cdot 10^{10}$		0.016	$1.1 \cdot 10^8$
Coarse-grained sandstone	0.34	$9.6 \cdot 10^7$		0	$1.8 \cdot 10^9$
	0.18	10^{10}	Olivene-pyroxenite	0.028	$0.7 \cdot 10^7$
Medium-grained sandstone	1.0	$4.2 \cdot 10^5$		0.014	$0.39 \cdot 10^8$
	1.67	$3.18 \cdot 10^8$		0	$0.56 \cdot 10^{10}$
	0.1	$1.4 \cdot 10^{10}$	Basalt	0.95	$4.1 \cdot 10^6$
Pyrophyllite	0.76	$6.1 \cdot 10^3$		0.49	$9.0 \cdot 10^7$
	0.72	$4.9 \cdot 10^9$		0.26	$3.1 \cdot 10^9$
	0.7	$2.1 \cdot 10^{10}$		0	$1.26 \cdot 10^{10}$
	0	$\sim 10^{13}$	Peridotite	0.1	$3.07 \cdot 10^5$
Granite	0.31	$4.4 \cdot 10^5$		0.003	$4.0 \cdot 10^5$
	0.19	$1.8 \cdot 10^8$		0	$6.5 \cdot 10^5$
	0.06	$1.3 \cdot 10^{10}$			
	0	10^{12}			

resistivity as water is removed probably is caused by the presence of significant amounts of metallic minerals.

The relationship between rock resistivity and water content is well-developed only over the range 0-2 to 4% water content. At higher water contents, the resistivity of a rock is less affected by changes in water content. The large effect small amounts of water have on conductivity may be explained as follows: the water forms thin, continuous films over the grains; as a result, we have a system in which water acts as a conductive matrix, and the resistant, rock-forming minerals are merely inclusions. Thus, according to Semenov's theory [91], a small increase in the conductive component must cause a large change in resistivity, as is observed experimentally.

Resistivity of Water-Saturated Clay-Free Rocks. Carbonate rocks, sandstones, and sand are included in this group of rocks. Because the rock framework in such rocks is made up of highly resistive mineral grains, the resistivity is

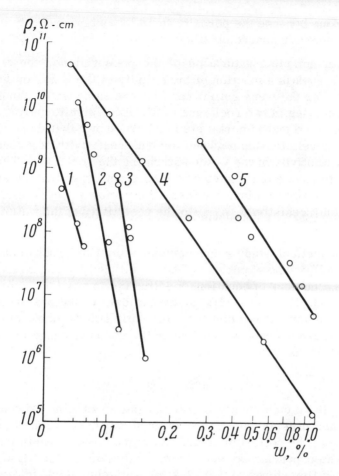

Fig. 29. Relationships between resistivity and moisture content: (1) andesite-basalt, (2,3) granite, (4) dolomite, (5) basalt.

determined mainly by the amount of water present, its salinity, and by the way the water is distributed through the rock.

The amount of water that can be held in a rock at full saturation is limited by the available porosity. The greater the porosity, k, the greater will be the water content, w. Therefore, for this group of rocks, the relationship between resistivity and porosity is the same as the relationship between resistivity and water content. Actually, in practice the relationship $\rho = f(k)$ is used.

This is done because the porosity is an important property describing the reservoir characteristics of a rock.

With complete saturation of the pore volume, the resistivity of a rock is a function of the salinity of the water in the pores. However, for the same salinity and for the same water contents, the resistivities of two rocks may still differ significantly. Both theoretical and experimental studies have shown that there should be a linear relationship between the bulk resistivity of a rock, ρ_{vp}, and the resistivity of the water saturating the rock, ρ_v. Therefore, in order to have a parameter which does not depend on the salinity of the saturating solution, the formation factor P_p is defined as the ratio of bulk resistivity, ρ_{vp}, to the resistivity of the water filling the pores, ρ_v.

Theoretical studies for various simplified rock models, such as collections of cubes, spheres, ellipsoids, and so on, have provided a number of equations relating the formation factor to porosity [86, 139, 140-144]. Mathematical evaluations of experimental data have shown that the most satisfactory expression $P_p = f(k)$ for an inhomogeneous rock over the porosity range from 3-5% to 20-40% as follows:

$$P_p = \frac{a_p}{k^m}, \tag{III.8}$$

where a_p is an empirically defined parameter ranging between 0.4 and 1.4 and m is an empirically defined exponent, related to the structure of the pore volume and the degree of cementation of the rock. The exponent may vary from a value of 1.3 for loose sands and oolitic limestones to $2.0-2.2$ for well-cemented, low-porosity sandstones.

In graphical presentations of the function $P_p = f(k)$, experimental data are usually grouped according to petrology, grain sizes, and the amount of cementation of the rock-forming minerals; grouping of experimental data by rock type only partially eliminates the effect of pore shape. As may be seen from Fig. 30, the porosity may vary widely in a number of rock types.

Based on a statistical study of a great volume of data reported by investigators both in the Soviet Union and in other countries for limestones and sandstones, Dakhnov [139] has constructed the average curves shown in Fig. 31, which are of the form of equation

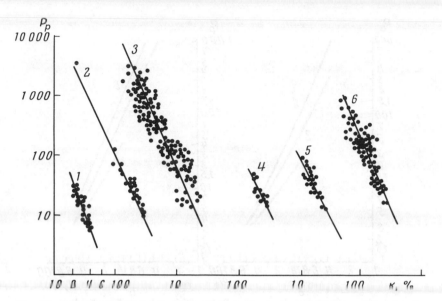

Fig. 30. Relationship of the formation factor, P_p, to the coefficient of porosity, $k_π$ for carbonate (limestone) sedimentary rocks in the USSR and the USA: (1) lime-stones from Kazakhstan (from Sigal), (2) limestones from the Kuibyshev area, Bashkiria (from Kachurina) (3) carboniferous limestones from the Saratov Basin (from Eidman), (4) devonian limestone from Crosset, Texas, (5) oolitic limestone, Smackover formation, (6) Permian limestone, San Andreas Basin, Texas (from Archie).

(III.8). Values for the parameters a_p and m for these curves are listed in Table 23.

These curves indicate that the greater the degree of cementation of a rock, the larger will be the formation factor, as well as resistivity, for a given water content. With increasing cementation, pore shape becomes progressively more important in determining the resistivity. With increasing cementation, the added material reduces the cross section of portions of the pore structure and increases the tortuosity. The tortuosity of a pore structure is defined as the statistical length of the pore structure between two parallel planes, in relation to the actual distance between these planes. The effect of tortuosity on the electrical conductivity of a rock is fundamental because two rocks saturated with the same amounts of a solution with the same salinity, but with different

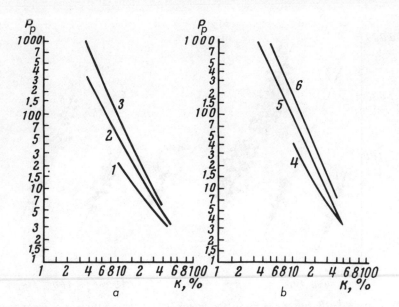

Fig. 31. Averaged relationships between formation factor P_p and the co-efficient of porosity (water content by volume) for sandstones and carbonate rocks (from (Dakhnov): a-sandy-shaly rocks, b-carbonate rocks. (1) Loose sand, (2) weakly cemented sand, (3) moderately cemented sandstone, (4) unconsolidated limestone grains, (5) coarsely crystalline dense limestone and dolomite of moderate density, (6) finely crystalline.

Table 23. Values a_p and m for Curves of the Form $P_p = a_p/k^m$

Curve number on Fig. 31	a_p	m
1	1	1.3
2	0.7	1.9
3	0.5	2.2
4	0.55	1.85
5	0.6	2.15
6	0.8	2.3

textures, may have quite a different resistivities [145-148]. The effect of pore geometry becomes particularly important in well-cemented rocks. As a result, in many rocks the correlation between resistivity and permeability is much better than the correlation between resistivity and porosity [149-152].

In cases in which the complete pore structure is saturated with an electrolyte and contributes to conduction, the formation factor is related to tortuosity, T, and porosity, k, by the equation [139]

$$P_p = \frac{T^2}{k}.$$

Published data are in good agreement with this expression. Knowing the resistivity of a rock from an electric log, and knowing the resistivity of the formation water, it is a simple matter to determine the porosity using the curves $P_p = f(k)$ for the appropriate rock type [143, 144]. The inverse problem may also be solved — that of estimating the resistivity of a rock when the porosity and water resistivity are known. In addition, knowledge of the relationship between formation factor and porosity has a further value, in that it may be possible to estimate the lithologic composition of a rock using resistivity data.

Shaly Rocks. In shaly rocks, in addition to nonconducting minerals, zeolites and clay minerals which serve as current conductors are found. As a result, the resistivity of a shaly rock is a function not only of the resistivity of the pore water, but also of the amount of clay present, its properties, and the manner in which it is distributed in the rock [153, 154]. Clay may occur as aggregates between grains in a rock, or it may occur as a coating over the grain surfaces. Significant amounts of clay in a rock lead to a number of peculiar effects. In the first place, with increasing clay content, the water saturation may exceed the porosity because of interactions between the clay minerals and the water [155]. Secondly, the linear relation between bulk resistivity and water resistivity no longer holds. A highly saline solution increases the formation factor of a shaly rock, while a dilute brine reduces the formation factor. Graphs of the relation between formation factor and brine salinity, given by Éidman [146], are shown in Fig. 32. A similar group of curves $P_p = f(C_1)$ for rocks with varying amounts of

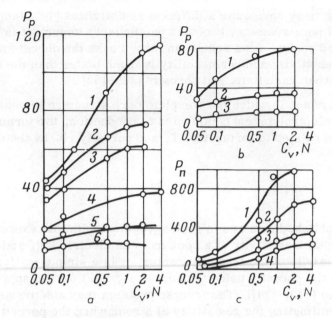

Fig. 32. Relationship between formation factor P_p and the concentration C_V of a solution of sodium chloride saturating shaly-sandy rocks: a — sandstones, (1, 2) shaly, (3, 4) slighly shaly or silty, (5, 6) clean; b — siltstones with varying degrees of shaliness; c — siltstones.

clay has been given by Vendel'shtein [156]. It is obvious that the effect — the change in formation factor with salinity – is more pronounced in clay-rich rocks. The effect of salinity on the formation factor of shaly rocks is accompanied by a variation in resistivity for rocks as a function of grain size. The resistivity of fine-grained rocks saturated with a high salinity brine is always greater than the resistivity of medium- or fine-grained rocks saturated with the same brine. With low salinities, the opposite is true.

These phenomena have been explained by Dakhnov as follows. For a highly saline electrolyte, a portion of the ions leave the solution and enter an electrical double layer. Ions in the double layer have a lower mobility than free ions in response to an externally applied electric field; therefore, the resistivity is increased. Thus, the greater the amount of clay in a rock, the larger will be the surface area over which the double layer forms, and the

greater will be the relative increase in the resistivity of the rock.
Lowering of the resistivity of shaly rocks when they are saturated
by dilute solutions is explained by the effects of surface conductance, which arises from partial hydrolysis of clay minerals. The
results of this hydrolysis contribute to conductivity. The higher
the shaliness, the greater will be the number of ions released in
hydrolysis and, therefore, the lower will be the resistivity of the
rock.

Surface conductivity depends on the existence of a high-con-
ductivity double layer, which becomes thinner with increasing salinity and finally becomes insignificant [157].

In considering the effect of surface conductance, a special
parameter related to the percent clay content in a rock is defined
by the ratio of the formation factor for a clayey rock saturated
with an electrolyte of a given salinity to the formation factor for
the same rock in a clay-free condition. The effect of shaliness is
to increase this parameter with increasing water resistivity. The
relationship between resistivity in shaly rocks and porosity is quite
complicated in detail [24, 139].

The presence of shaly material in oil-bearing rocks reduces
their resistivities markedly, so that the resistivities of such rocks
may be moderate or low.

Electrical Resistivity of Oil-Bearing Rocks

It has been indicated in the preceding section that the resistivity of a rock which is completely saturated with a brine is determined by the porosity to a first approximation. However, the pore
spaces in a rock may not always be saturated with an aqueous electrolyte. Sedimentary rocks above the water table or forming gas
reservoirs may be only partially saturated with water. Also, in
oil reservoirs, oil may partially replace water in the pore spaces.
The presence of oil, natural gas, or air in the pore structure of
a rock may increase the resistivity significantly over what it would
be in a completely water-saturated rock. Dry rock, oil, and gas
have practically infinite resistivities. The higher the oil or gas
saturation in a rock, the higher will be the resistivity. However,
for a given oil or gas saturation, the resistivity of a rock may
vary widely, depending on the salinity of the residual water in the
pores, the distribution of fluids in the pores, and so on.

A dimensionless parameter, the resistivity index, P_n, which is similar in nature to the formation factor is defined so that the effects of these factors may be studied, particularly in reservoir rocks. It is a measure of the ratio by which the resistivity, ρ_{np}, of a partially saturated rock is increased over the resistivity, ρ_{vp}, of the same rock when completely water saturated

$$P_n = \frac{\rho_{np}}{\rho_{vp}}.$$

The value for the resistivity index is a measure of the oil or gas saturation in a rock. If we assume that there are no fluids except water and oil in the pores of a rock, and if the fraction of the pore space saturated with water is designed as k_v and the fraction saturated with oil as k_n, the sum of these two coefficients will be unity. According to experimental data published by Dakhnov [139], the relationship between the resistivity of a rock and its oil saturation k_n and water saturation k_v is given by

$$P_n = \frac{\rho_{np}}{\rho_{vp}} = \frac{a_n}{(1 - k_n)^n} = \frac{a_n}{k_v^n}. \tag{III.9}$$

This expression is valid providing the parameters are assigned the following values: a) in sandstones and shales (for $k_v < 40\%$), $a_n = 0.6$ and $n = 2.25$; b) in carbonate rocks (for $k_v < 25\%$), $a_n = 0.4$ and $n = 2.1$.

Guyod and Archie have given another expression for the relationship $P_n = f(k_v)$, as follows:

$$P_n = k_v^{-n}.$$

According to Guyod, $n = 1.98$, and according to Archie, $n = 2$.

With increasing oil saturation, the resistivity index and the rock resistivity both increase. A number of studies of the behavior of the resistivity index in shaly rocks have indicated that commercial quantities of oil may be produced from such rocks at lower values for the resistivity index than would be required in clay-free rocks. because shaliness reduces the resistivity index.

It should be noted that experimental data for the function $P_n = f(k_v)$ may not always be described accurately with the expressions given earlier. To some degree, this is explained because equation (III.9) does not take into consideration the fabric of the

solid framework of a rock or the manner in which fluids are distributed through the pore space. The distribution of the water in
the pores depends on the wettability of the mineral grains. In
most cases, mineral surfaces are hydrophilic, and a thin coating
of water covers all grains. However, if minerals are hydrophobic,
they will be coated with films of oil, and the water will assume the
form of isolated droplets insulated from each other. As a result,
the resistivity of a hydrophobic rock with a given water saturation
is much greater than that of a hydrophilic rock [158]. This effect
is illustrated the the curves in Fig. 33.

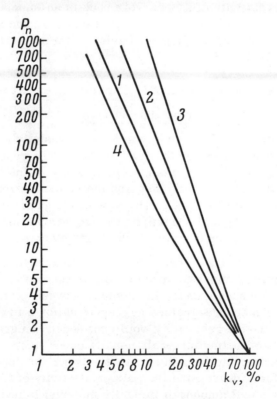

Fig. 33. Relationship of the resistivity index, P_n,
to the coefficient of water saturation, k_v: (1, 2, 3)
sandy-shaly rocks which are hydrophilic, moderately
hydrophilic, and hydrophobic, respectively, (4) carbonate rocks (from Archie).

For the same oil saturation, the resistivity index (and the resistivity, as well) for a hydrophobic rock is nearly an order of magnitude larger than the index for a hydrophilic rock. Rocks containing oil–water droplets are dispersed in oil and cannot serve effectively in conducting current.

Kotov [159] in experiments with sandstone samples has found the oil–water emulsions can form in the presence of large-molecule naphtha acids. In order to avoid errors in the laboratory determination of the function $P_n = f(k_V)$, it is recommended that measurements not be made on samples originally saturated with oil and water in the reservoir under investigation.

The resistivity of a reservoir rock is an important criterion in estimating the oil saturation in clastic rocks, and it is widely used in reservoir engineering as a basic tool for predicting productivity [160, 161].

Relationship Between Electrical Resistivity, Texture, and Anisotropy in Rocks

The effect of fabric or texture on the resistivity of various rocks varies depending on the type of rock. The resistivity of an ore-bearing rock depends mainly on the shape of the ore-mineral grains and the way in which they are distributed through the host rock, as was stated earlier. The size of the grains is less important than the shape, but variations in size account for differences in resistivity between two rocks which are otherwise identical.

Let us consider a rock which would be composed of grains of a low-resistivity mineral, in which there were no nonconducting impurities. In such a case, the contact between grains would be the same for a coarse-grained aggregate as for a fine-grained aggregate, and the resistivity would not depend on grain size. In reality, nonconducting minerals usually occur in ores, as well as in ore minerals, blocking the contact between the conducting grains and, therefore, decreasing the bulk conductivity of the rock. Only a relatively small amount of impurity material is necessary to surround each conducting grain with a monomolecular film. The question of the relation between the volume of intercrystalline films and the shape and size of grains in the aggregate has been examined by Kazer [in 91]. He obtained the following relationship between grain diameter and the amount of impurity required to form a monomolecular film around grains:

Grain diameter in mm.	1	0.1	0.01
Amount of impurity	0.0001%	0.001%	0.01%

These data indicate that the finer the sizes in a polycrystalline aggregate, the greater will be the amount of impurity required to form an insulating film around the grains. Considering these data, we may conclude that rocks with identical compositions, fabrics, and amounts of impurities but which differ in grain size will also differ in resistivity. The coarser the grain size in the rock, the higher will be the resistivity, all other things being equal.

The effect of grain size on the resistivity of highly resistive rocks has not been studied, but we may suppose that there is an effect. It is obvious that if the conducting minerals can form continuous filaments, or if an aqueous phase is present, the grain size may play an important role in determining the resistivity. Thus, the tortuosity of conductive paths should be less in coarse-grained rocks, and the amount of conductive material required to form conducting filaments will be less. One would suppose that the inverse type of relationship between conductivity and grain size would hold for a rock in which conductive impurities were distributed around insulating grains rather than insulating around conducting grains.

The best defined relation between resistivity and grain size is found in sedimentary rocks. As stated earlier, the resistivity of a fine-grained rock saturated with a low-salinity brine is less than the resistivity of a medium- or fine-grained rock saturated with the same amount of the same brine. The inverse relation is found for rocks saturated with high-salinity brines. The effect of grain size on the resistivity of a sedimentary rock may be explained in part by the fact that tortuosity is greater in fine-grained rocks.

The resistivity in nonmineralized rocks depends not only on grain size, but also on grain shape and fabric, or texture. In addition, mineral grains may be preferentially oriented with respect to their internal crystal structure. Such orientation, as well as other types of preferential orientation, leads to anisotropy in resistivity.

Table 23a. Resistivity of Anisotropic Rocks

Rock type	Resistivity, Ω-cm		Reference
	transverse to layering	parallel to layering	
Ore: zinc blend and galena	$3.6 \cdot 10^4$	0.1	[22]
Crystalline slate with sericite	$7 \cdot 10^6 - 3 \cdot 10^9$	$6 \cdot 10^6 - 5 \cdot 10^7$	[22]
Shaly slate	10^7	$5 \cdot 10^4$	[22]
Anthracite	$4.5 \cdot 10^3$	$7.9 \cdot 10^2$	[128]
Coal	$1.71 \cdot 10^3$	$0.7 \cdot 10^3$	[128]

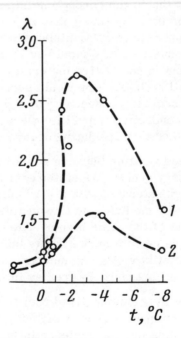

Fig. 34. Variation of the coefficient of anisotropy for a laminated
shale as a function of temperature:
(1) sample of shale with thin interlayers of sand, (2) laminated clay.

There are some rocks in which microanisotropy is developed in the same manner as macroanisotropy, through the alternation of fine layers with different petrographic properties. In such a case, the anisotropy of the complex of layers will be greater than the anisotropy for the individual layers [139, 162]. Examples may be found in sedimentary rocks which consists of alternating layers of dense rock with high resistivity and less dense rock with lower resistivity (shale layers, sandstone layers, and so on).

The resistivities along and perpendicular to the bedding planes are given by [139]

$$\rho_t = \frac{(\nu + 1)\,\rho_p \rho_s}{\nu \rho_s - \rho_p}$$

and

$$\rho_n = \frac{\nu \rho_p + \rho_s}{\nu + 1},$$

where ρ_p is the resistivity of the poorly conducting interlayers, ρ_s is the resistivity of the highly conducting layers, and ν is the ratio of the total thickness of layers with resistivity ρ_p to the total thickness of layers with resistivity ρ_s.

The coefficient of anisotropy in resistivity, designated as λ, is:

$$\lambda = \sqrt{\frac{\rho_n}{\rho_t}} = \sqrt{1 + \frac{\nu}{(\nu + 1)^2}\,\frac{(\rho_p - \rho_s)^2}{\rho_p \cdot \rho_s}}.$$

The resistivity transverse to the layering is always larger than the resistivity along the bedding. Experimental data show that the anisotropy increases with increasing contrast between the resistivities of the various interlayers. Values for ρ_n and ρ_t for various anisotropic rocks are listed in Table 23a.

Anisotropy may account in part for the differences between resistivities measured on rocks in place and in the laboratory.

It should be noted that there is a marked dependence of the coefficient of anisotropy, λ, on temperature (Fig. 34). As may be seen from this illustration, the coefficient of anisotropy for λ laminated clay varies from 1.2 to 2.75 and then decreases again over the temperature range from 0 to -2.5°C. This peculiar

behavior of λ as a function of temperature near the freezing point
has been explained by Ananyan and Dobrovol'skii [163] as being
caused by the sequential conversion of disoriented, weakly oriented,
and oriented water molecules to ice. The maximum value for λ is
associated with the temperature at which the first two types of
water have frozen and the remaining water is in layers adsorbed
on the grains. The formation of interlayers of ice markedly increases the resistivity ρ_n; with further lowering of temperature,
the rock becomes more homogeneous, with an increase in ρ_t. Thus,
as temperature is further decreased, the coefficient of anisotropy
decreases toward some constant value.

Anisotropy in rocks may be caused by preferential orientations of joint patterns, as well as by the factors discussed in the
preceding paragraphs [164-167].

In reference [167] it has been shown that rocks commonly
contain two main families of vertical joints, usually perpendicular
to one another. It some cases, these vertical joints may be
supplemented by a diagonal family of joints. Also, there are cases
in which a rock has only one horizontal joint.

The various cases, classified using Nechai's system, are
illustrated in Fig. 35, and the formulas for computing resistivity
for each of the cases are given in Table 24 [165].

The following symbols are used in Table 24: ρ_0 is the resistivity of a rock along the coordinate axis, ρ_p is the resistivity of
the nonjoined rock fragments, ρ_v is the resistivity of the water
filling the joints, and m is the joint porosity. Very good agreement has been obtained between resistivity computed with these
formulas and resistivities measured on joint samples [165, 168].

Goryunov [164] has considered the resistivity of idealized
jointed rocks, considering not only the relative distribution of
joints, but also their character. For rocks with two mutually perpendicular sets of joints, he found different values for the coefficient of anisoptropy, depending on the width of the joints and the
ratio of the resistivity of the unfractured rock to the resistivity of
the material filling the joints. The greater the value for this ratio,
the more rapidly does the coefficient of anisotropy increase with
joint width.

The reservoir properties of rocks are determined not only by
the availability of primary porosity, but also to a large degree by

Table 24. Formulas for Computing Resistivity in Jointed Rocks [165]

Case	Axis along which current flows	Formula for bulk resistivity	
1B 1G	$\left.\begin{array}{c}X\\Z\end{array}\right\}$	$\rho_0 = \rho_p - m(\rho_p - \rho_v)$	(1)
1B 1G 2B 2GB	$\left.\begin{array}{c}Z, Y\\X, Y\\Z\\Y\end{array}\right\}$	$\dfrac{1}{\rho_0} = \dfrac{m}{\rho_v} + \dfrac{1-m}{\rho_p}$	(2)
2B 2GB	X, Y X, Z	$\dfrac{1}{\rho_0} = \dfrac{m}{2\rho_v} + \dfrac{1}{\rho_p}$	(3)
3 X	X, Y, Z X, Y, Z	$\dfrac{1}{\rho_0} = \dfrac{2m}{3\rho_v} + \dfrac{3-2m}{\rho_p(3-m)}$	(4)

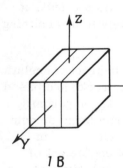

1B

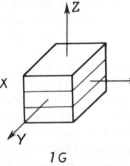

1G

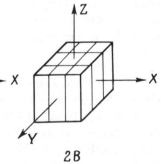

2B

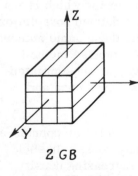

2 GB

3

X

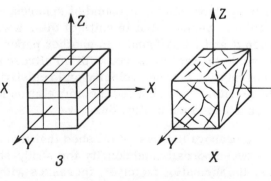

Fig. 35. Schematic representation of various forms of fracturing in rocks.

the presence of widely distributed small fractures. Therefore, there is considerable value in studying the resistivity of jointed rocks as a function of direction. Moreover, data on the coefficient of anisotropy in electrical resistivity allows determination of heterogeneity in interbedding in sedimentary rocks, the degree of metamorphism, and the jointing of rocks, among other things.

Relationship Between Resistivity and the Other Physical Properties of Rocks

There is considerable scientific and practical interest in the study of correlations between the electrical properties of rocks and the magnetic and elastic properties. The properties are basic to the applications of two widely used geophysical exploration methods – magnetic and seismic. In field surveys, frequently it is necessary to have data on the correlations between properties in interpreting the results of surveys made with several methods.

Moreover, there is a far greater volume of data available from electric logging than from acoustic logging. As a result, it is desirable to use electrical data rather extensively in determining acoustic properties.

Suppositions concerning the existence of an indirect correlation between elastic and electrical properties in clastic rocks are based on the dependence of both types of properties directly on porosity or density. Polak and Rapoport [169], in analyzing a large amount of experimental data, have found an increase in velocity, v, in sandy-shaly rocks with increasing resistivity, the coefficient of correlation being 0.65. It should be noted that the velocity was determined from elastic rebound of spheres, a method which is not particulary precise. But in spite of this, a well defined correlation between ρ and v was found. In another paper by these same authors [170] velocity was measured with an ultrasonic impulse method. The correlation between velocity and resistivity was substantiated by these data. Data showing the correlation for sandstones and siltstones are given in Fig. 36.

Kobranova [24] has established that there is a linear correlation between porosity and density for sandy-shaly rocks. In such rocks, the formation factor P_p increases with increasing density

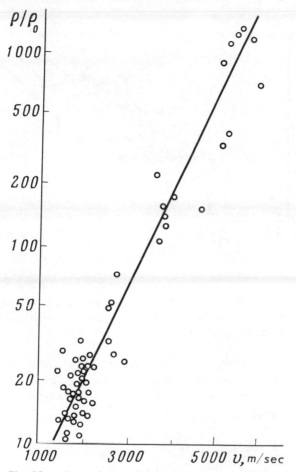

Fig. 36. Relation between the relative resistivity of sandstones and siltstones and the velocity of propagation of
acoustic waves.

(Fig. 37)*. Ukleba [171] has attempted to determine the strength
of rocks from measurements of resistivity. The existence of such
a correlation is based on the supposition that the strength of a rock
is reduced by the filtration of liquids between the grains at the same
time the resistivity is reduced. A relationship is apparent in com-
parisons of data on the resistivities and strengths of rocks – resistivity increases with increasing strength.

*The formation factor, which is directly related to rock resistivity, will be used inter-
changeably with resistivity in the following sections.

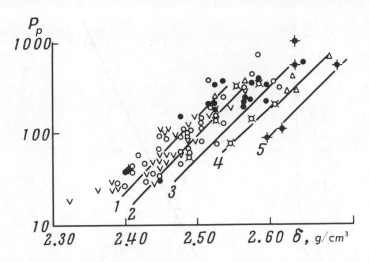

Fig. 37. Relationship between formation factor P_p and density δ for sandy-shaly layers in Bashkiria and Tataria. Resistivity would be expected to vary along the lines shown for constant densities, g/cm³, as follows: (1) 2.64-2.65, (2) 2.66-2.67, (3) 2.68-2.69, (4) 2.70-2.71, (5) 2.72-2.74.

It appears that studies of correlations between electrical properties and density are still in a very early stage and we do not know of any studies of a possible correlation between electrical and magnetic properties.

The Effect of Temperature on the Resistivity of Rocks

The temperature range over which rock temperatures may occur is measured in hundreds of degrees. In the far north in regions of permafrost, rock temperatures go as low as -40°C, while at the boundary between the core and mantle, the rock temperature may be as high as 700°C [172]. Data on the resistivity of rocks at relatively low temperature (-40° to +20°C) are important in geophysical exploration for active and frozen layers in permafrost zones. Somewhat higher temperatures are of concern in electrical well logging. It is necessary to know the relationship between resistivity and temperature for rocks and minerals to temperatures of 1500°C in interpreting deep electromagnetic sounding data.

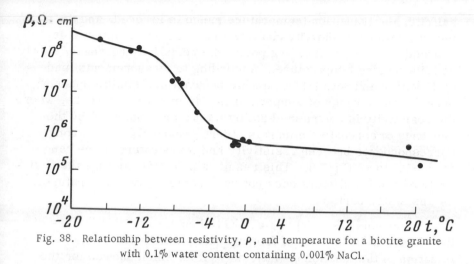

Fig. 38. Relationship between resistivity, ρ, and temperature for a biotite granite with 0.1% water content containing 0.001% NaCl.

The study of rock resistivity at very low temperatures and very high temperatures is and has been undertaken by a relatively small group of investigators. References [173-175] describe the earliest work on measuring resistivities at very low temperatures. The basis for these studies was the determination of various permafrost conditions in exploration. The authors noted the behavior of earth resistivity and its variation with temperature close to 0°C. A good review of this work is included in a paper by Dostovalov [33]. Studies of the resistivities of rocks at low temperatures have been reported by Nesterov and Nesterova [176], Grutman [177], and Ananyan [178, 179].

Reference [176] describes the first study of the resistivity of various rock types at low temperatures. The paper describes the results of measurements over the temperature range -20° to 20°C for rocks such as biotite granite, porphyry, diabase, chalcopyrite, and various sandstones. It was established that the resistivity-temperature function was of the same form for all of the rocks studied except chalcopyrite (see Fig. 38). In increasing temperature from -20° to -10° or -8°C, the resistivity was found to decrease slowly. At slightly higher temperatures, the resistivity was found to decrease rapidly with increasing temperature as a result of a change of state from ice to water. The limits for this transition range of temperatures was found to depend on the salinity of the water in the rock, to some degree. With increasing pore-water

salinity, the transition temperature range is lowered, and the a-
mount by which resistivity changes in the transition interval is
reduced. Above the freezing point, the resistivity decreases slowly
with increasing temperature. According to measurements made
by Mikailov and Soya [180], marble behaves in a similar manner
over the same range of temperatures. In mineralized rocks, where
the resistivity is determined not by the water content but by the
presence of conductive minerals, there apparently is no strong
correspondence between resistivity and temperature in the range
from -19° to +8°C [176]. This result is plausible because the resistivity of semiconductors does not vary rapidly with temperature,
as was stated earlier.

Ananyan [178, 179] has conducted comprehensive studies of
the electrical resistivity of frozen rocks, which allowed a determination of the function $\sigma = f(t)$, as well as which showed for the
first time the existence of a maximum in electrical conductivity as
a function of the water content in frozen rocks, along with a theoretical justification. In working with sedimentary rocks with various textures, Ananyan came to the conclusion that at temperatures
near zero the behavior of resistivity depends on the structure of
the water contained in the pore spaces. Water may be present in
rocks in a crystalline form, as adsorbed water and as free water.
In considering the behavior of resistivity, only the relative amounts
of adsorbed and free water need be considered (the water which
may form an electrolyte).

The unique properties of adsorbed water appear to explain
the difference in the function $\sigma = f(t)$ for coarse-grained and fine-
grained rocks [178, 179]. In coarse textured materials such as
sands and gravels, only an insignificant amount of adsorbed water
occurs and, as a result, the change of state between water and ice
takes place close to 0°C. The resistivity-temperature curve in
these materials is characterized by a sudden jump at the freezing
point. In fine-textured rocks with large specific surface areas,
there will be a high fraction of adsorbed water. As a result, the
change of state between water and ice takes place below 0°C. On
freezing, first the free water is converted to ice, and then the
weakly adsorbed water is converted. Thus, the rate of change of
resistivity with temperature $\Delta\rho/\Delta t$ on freezing will be lower for
fine-textured rocks than for coarse-textured rocks. Moreover,
these fine-grained rocks, because of their low ice content, may

have a lower resistivity than coarse-textured rocks above 0°C. The resistivity of frozen rocks depends to a large extent on their water content.

At positive temperatures, resistivity decreases with increasing water content. On the other hand, in frozen rocks, Ananyan and Dobrovol'skii [163] have found the inverse relationship — rocks with the most water have the highest resistivity. This may be explained by the fact that the amount of water converted to ice is proportionally greater in rocks with a high water content. Experimental data showing the higher resistivity in rocks with large contents and the rapid variation of resistivity with temperature in such rocks are summarized in Fig. 39.

Moreover, some frozen fine-textured rocks exhibit a maximum in the electrical conductivity curve depending on water content. The maximum in electrical conductivity at negative temperatures is associated with water contents near those which cause plasticity in clay [163, 178]. This phenomenon is well-illustrated by experimental data obtained with clay and shown in Fig. 40. Comparing these curves at subzero temperature, it is obvious that the highest conductivity in the frozen state is associated with a water content of 24.0% (curve 2 assumes the highest position on the graph). According to Ananyan, this particular water content is approximately that needed to cause plasticity. These data suggest that at water contents below the amount needed to make a clay plastic, the water is present entirely in an adsorbed state. As a result, the

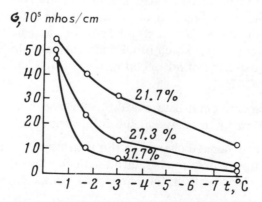

Fig. 39. Curves giving the relationship between conductivity and temperature for frozen clays with different water contents.

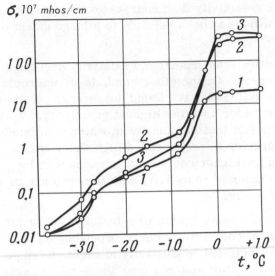

Fig. 40. Relationship between conductivity and temperature for clays with various water contents: (1) 9.3%, (2) 24.0%, (3) 45.6%.

conductivity in such rocks is significantly higher than that in rocks with either more or less than the critical water content for plasticity. In one case, there is a large fraction of free water which is quickly converted to ice on freezing, reducing the conductivity, and, in the other case, there is little water present in the rock to form conducting paths, even though it is all adsorbed, and the resistivity is high. The conditions for maximum current conduction are the absence of ice crystals, which block current, and the presence of significant quantities of adsorbed water, which has a relatively good conductivity [157].

The experimental data available for rocks at sub-zero temperatures allow us to arrive at the following conclusions:

1. A pronounced change in the resistivity as a function of temperature is associated with the freezing or thawing of an electrolyte in the pores of a rock.

2. The actual resistivity at negative temperature depends on the water content, the chemical composition of the electrolyte, and the grain size of the rock.

3. The resistivity of fine-grained rocks (shale) increases less abruptly with decreasing temperatures than does the resistivity of coarse-grained rocks.

4. With increasing salinity of the pore water, the increase in resistivity on freezing is reduced.

Data on the relationship between resistivity and ice content at freezing temperatures may be used in interpreting electrical geophysical surveys made in permafrost areas.

The relationship $\rho = f(t)$ for rocks over the temperature range from 20 to 200°C also has a number of peculiarities, depending on water content. In considering the resistivity of water-bearing rocks in this temperature range, we must examine the summed effects of changes in the liquid and solid fractions.

The effect of temperature on the resistivity of a solution is given by the following formula:

$$\rho_{vt} = \frac{\rho_{v18}}{1 + a_t (t - 18°)} = P_t \rho_{v18},$$

where ρ_{vt} is the resistivity of the solution at the ambient temperature t°C, ρ_{v18} is the resistivity of the solution at a reference temperature of 18°C, and α_t is the temperature coefficient for electrical resistivity, which averages about 0.025/°C. The value of α_t increases with decreasing salinity. The formula shows that the resistivity of an electrolyte decreases with increasing temperature.

The rate of change of resistivity with temperature persists to temperatures up to 150°C (see Fig. 41). At higher temperatures, and for low-salinity solutions, the rate of decrease in resistivity with temperature diminishes, and, according to Polyakov [183], for solutions with a salinity of less than 1%, the resistivity increases with increasing temperature. Because sedimentary rocks lie in the upper part of the earth's crust, where the temperature does not exceed 200 to 250°C, a decrease in resistivity with depth caused by the decrease in the resistivity of the pore water with temperature is observed.

The variation of the conductivity of the solid minerals in a rock with temperature is given by one of the two following formulas, depending on whether the minerals are electron semiconductors or dielectric insulators

$$\sigma_t = \sigma_0 e^{-\frac{E_0}{2kT}}$$

and   (III.10)

$$\sigma_t = \sigma_0 e^{-\frac{E_0}{kT}} .$$

According to these equations, the conductivity of the solid phase also increases with increasing temperature.

Depending on the relative proportions of liquid and solid fractions in a rock, the bulk resistivity will vary according to the expression for a solution or according to the expression for solids. If there is a significant amount of water in a rock, and if the water content does not change with temperature, the effect of increasing temperature, $\rho = f(t)$, will be that for a solution. However, if the rise in temperature causes an evolution of water from the rock, the resistivity will change anomalously. Studies of the temperature-resistivity curve for various types of coal, as well as various rocks, which have been made under laboratory conditions indicate that evolution of water is accompanied by a characteristic change in resistivity. The greater the original water content of a rock,

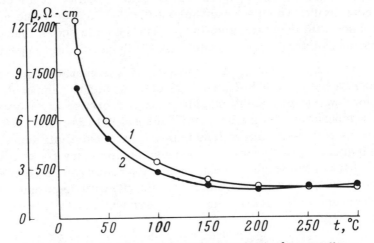

Fig. 41. Relationship between resistivity and pressure for two different concentrations of NaCl brine at 100 atm pressure: (1) 0.03%, (2) 10%, (from Polyakov).

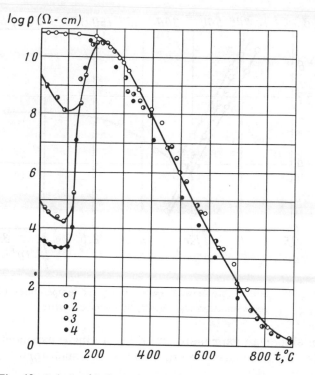

Fig. 42. Relationship between resistivity and temperature in
lower Moscovian coal saturated with a 10% NaCl brine.
Moisture content in percent: (1) 0, (2) 8.18, (3) 14.46, (4)
20.24.

the greater will be the change in resistivity on evolution. If the
initial water content is small, the decrease in resistivity of the
solid phase with increasing temperature may override the increase
in resistivity caused by evolution of the water. Data demonstrating
the change in resistivity for coals with various amounts of water
are shown in Fig. 42. In studying the water content in sandstones
and volcanic rocks, we have also found an ore increases resistivity as a function of temperature (see Fig. 43) and, in the best
examples, the resistivity remains nearly constant [133, 181] over
a range in temperatures.

In dry rocks without fractures, the relationship between resistivity and temperature over the range 100 to 700°C is well described by equation (III.10). When the curve log $\sigma = f(1/T)$ has a

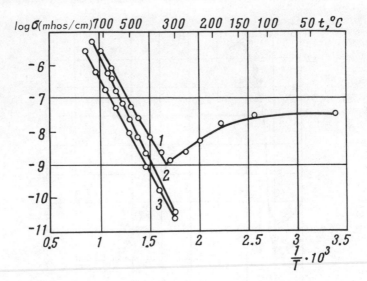

Fig. 43. Relationship between conductivity and temperature: (1) room-moist, sandstone, (2, 3) dry pyrophyllite.

change in slope; an expression with but a single exponential term (III.10) is not adequate, and the sum of two such terms is better:

$$\sigma_t = \sigma_1 e^{-\frac{E_0'}{kT_1}} + \sigma_2 e^{-\frac{E_0''}{kT_2}}.$$

The first term corresponds to the mechanism of conduction in the low-temperature region, while the second term corresponds to the mechanism of conduction in the high-temperature region. If there are more than two conduction processes in a given case, then more exponential terms must be included. Each term corresponds to a single conduction mechanism.

We shall now consider the results of studies of electrical conductivity for various types of rocks at high temperatures [181-192].

Study of the relationship between resistivity and temperatures above 200°C in sedimentary rocks is of no interest in geophysics, therefore, very little data has been reported. On the other hand, sedimentary rocks such as pyrophyllite, talc, and lithographstone have been used as coupling materials in equipment for studing electrical properties at high temperatures and pressures. I have

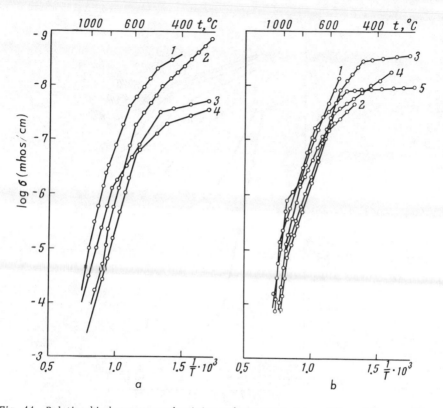

Fig. 44. Relationship between conductivity and temperature in acidic and intermediate rocks: a — (1, 2) andesite, (3) quartz diorite, (4) andesite basalt; b — (1) quartz, (2, 3, 4) granites, (5) perthite.

studied the electrical conductivity of pyrophyllite at temperatures from 250° to 900°C. The experimental data shown in Fig. 43 indicate that there is no change in conduction mechanism for this type of rock over the entire range of temperatures, and that even at 900°C the resistivity is still relatively high.

The temperature-induced variation of resistivity in igneous rocks has been studied by a number of investigators, including the author. The majority of measurements of the function $\log \sigma = f(1/T)$ have been made on basic and ultrabasic rocks. Acidic and intermediate rocks have been studied mainly by Noritomi and Asada [185, 186] and particularly in references [126, 182, 189, 190]. According to Marinin [126] and to Coster [182], the conductivity

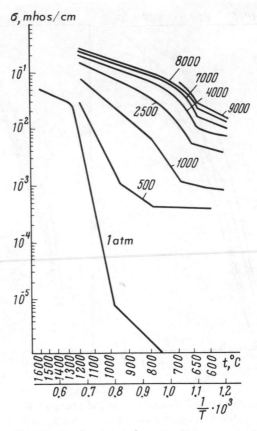

Fig. 45. Conductivity of granite during
heating to 1500°C at various pressures.

of granite and gneiss may be represented by single-segment lines
plotted to the coordinate system log σ and 1/T for temperatures up
to 750°C. Noritomi and Asada [185] have obtained a somewhat different character for the relationship log σ = f(1/T). The curves for
this relationship for acidic and intermediate rocks are shown in
Fig. 44, a and b, along with curves for the minerals quartz and
perthite. As may be seen, each of the resistivity-temperature
curves has several discontinuities in slope, indicating change in
the conduction process. The first discontinuity is observed at relatively low temperatures (450 to 550°C).

Table 25. Values for the Activation Energy, E_0, in Electrons Volts and the Logarithm for the Conduction Coefficient, σ_0, in mhos/cm for Acidic and Intermediate Rocks

Rock	Low-temperature region			High-temperature region			Reference
	Temperature range, °C	E_0	$\log \sigma_0$	Temperature range, °C	E_0	$\log \sigma_0$	
Perthite	540—720	1.3	—0.6	720—810	1.6	1.1	[185]
				810—980	0.7	—2.4	
				980—1200	5.6		
Quartz				570—780	1.2*	—1.0*	
				780—830	1.6	0.2*	[185]
				830—970	0.6*	—3.0*	
				970—1050	4.7	10.0*	
Granite	450—550	0.9	2.4	550—980	1.4	0.8	
				980—1200	4.5	14.0*	[185]
"				750—980	1.0	—1.0	[190]
				980—1200	4.85	>10	
"		0.6		200—980	0.62	4.45	[182]
				980—1100	2.5	5.0	
Diorite	200—600	0.86	—1.0	600—820	1.8	4.4	Author
				820—1000	1.0	—0.5	
Andesite	380—480	0.7	—2.2	480—800	1.4	1.1	[185]
				800—1100	1.6	2.3*	

*Data computed by É. I. Parkhomenko.

The last change in activation energy for charge carriers occurs in the temperature interval 800 to 1000°C. Lebedev and Khitarov [190], in studing dry granite, suggest the discontinuity at 970°C corresponds to the beginning of melting, which persists over a temperature range from 970 to 1250° (see Fig. 45).

Graphs such as those in Fig. 44 and 45 allow the determination of activation energy by noting the slope, $\tan \varphi$, of the curves $\log \sigma_0 = f(1/T)$, and substituting it in the formula $E_0 = 0.2 \tan \varphi$ to determine the constant σ_0.

Values for the activation energy, E_0, and the conduction coefficient, σ_0, as compiled by a number of authors, are listed in Table 25.

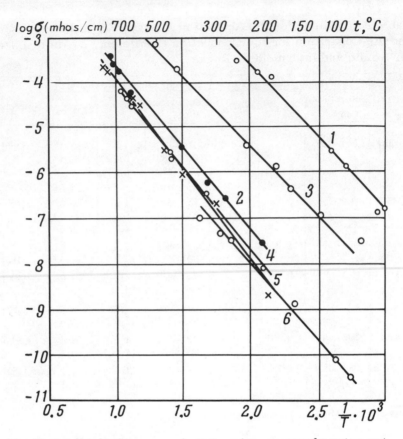

Fig. 46. Relationship between conductivity and temperature for various rock
samples: (1) peridotite, (2, 4) basalt, (3, 5, 6) diabase.

    Comparison of the values E_0 and σ_0 for the various rocks
indicates that the activation energy for charge carriers at temperatures below 600 to 700° is rather low, being 0.7 to 0.9 eV. At higher temperatures the activation energy is larger, assuming maximum values in the temperature range 1000 to 1200°C of 4 to 5.6 eV. The value for the coefficient σ_0 also increases as the activation energy increases. At temperatures above 1250°C, a drop in activation energy is noted which must be associated with melting. The existence of a low-temperature domain (below 450°C) characterized by low activation energy has been observed both by Noritomi and by the present author. However, it should be noted that residual water may play a role in conduction at these temperatures.

Basic and ultrabasic rocks are also characterized by changes in the mechanism of conduction with increasing temperature. A series of graphs for the function log σ =f(1/T) for such rocks, as reported by various authors, is given in Fig. 46-49. A number of studies, conducted jointly by Bondarenko and me [181], have indicated that there is no change in the conduction process in the following rocks in the temperature range from 100 to 500-800°C: diabase, basalt, and peridotite. The activation energy for these rocks in this temperature range does not exceed 0.8 eV, varying between 0.6 and 0.8 eV. At higher temperatures, Coster [182] has found a discontinuity in slope in the log σ =f(1/T) curve for gabbro, basalt, peridotite, and one sample of eclogite.

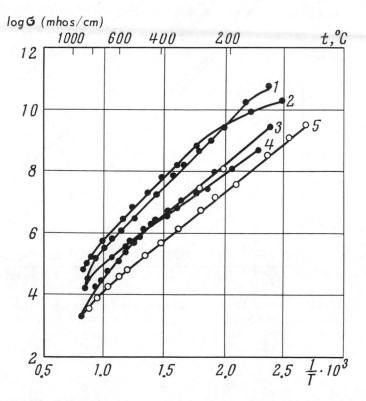

Fig. 47. Relationship between conductivity and temperature for rock samples [182]: (1) peridotite, (2, 3) eclogite, (4) gabbro, (5) crystalline basalt.

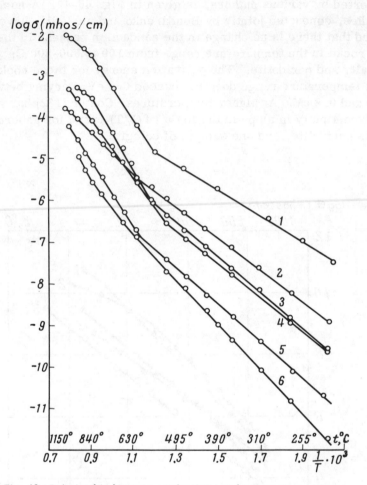

Fig. 48. Relationship between conductivity and temperature in igneous rocks: (1) andesite basalt, (2) norite gabbro, (3) diorite, (4) peridotite, (5) olivenite, (6) pyroxenite.

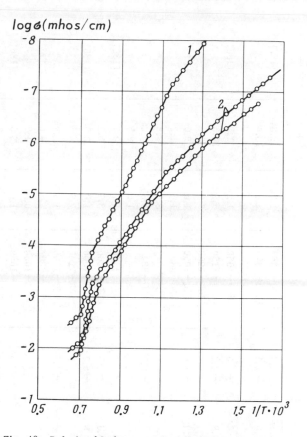

Fig. 49. Relationship between conductivity and temperature for olivene: (1) Sado deposit, (2) Ogazavara (Japan).

Noritomi [186], in the case of serpentinite and olivine, and
the present author, in the cases of diabase, pyroxenite, olivinite,
peridotite, and andesite, have found well-expressed discontinuities
in slope in the curve log $\sigma = f(1/T)$. Values for the activation energies and for the coefficient σ_0 for these rocks are listed in Table 26.

We may see from the data in Table 26 that the activation energy for basic and ultrabasic rocks below 650°C is 0.6 to 0.9 eV,
which is about the same as that for acidic and intermediate rocks
in many cases. Above 650°C, the activation energy increases to 1.6
eV and, in rare cases, to higher values, 4.6 to 5.2 eV for temperatures above 1050°C. A decrease in activation energy above 870°C
is found for diabase, and this may be associated with melting.

Table 26. Values for Activation Energy, E_0, in Electron Volts and the Coefficient, $\log \sigma_0$, in mhos/cm for Basic and Ultrabasic Rocks

Rock	Low-temperature region			High-temperature region			Reference
	Temperature range, °C	E_0	$\log \sigma_0$	Temperature range, °C	E_0	$\log \sigma_0$	
Olivene	300−460	0.66	−3.8	620−1060	1.64	2.34	[186]
"	460−620	1.0	−1.4	1060−1170	4.6	13.4	
"				>1170	0.7	−0.2	
"	300−480	0.64	−2.0	640−1050	1.2	1.3	[186]
"	480−640	0.8	−0.9	1050−1150	5.2	14.5	
"				1150−1200	0.8	0.9	
Olivenite	200−590	0.9	−1.6	590−1000	1.6	2.0	*
Olivene pyroxenite				500−860	1.4	1.2	*
"				>860	3.5	9.8	
Pyroxenite	200−630	1.1	−0.75	630−830	1.35	0.25	*
"				>830	1.5	2.5	
Serpentinite	300−500	0.4		630−1100	1.2	−0.7	[186]
"	500−630	0.6	−4.0	1100−1200	3.0	5.7	
"	300−600	0.5	−5.0	600−1000	1.6	0.9	[186]
"				1100−1200	2.6	4.5	
Peridotite	200−520	0.8	−1.5	520−1000	1.42	2.25	*
Andesite basalt	200−590	0.6	−1.2	590−830	1.6	4.8	*
"				830−1050	0.8	1.2	
"	300−450	0.1	−6.5	580−1050	1.6	3.5	[186]
"	450−580	0.3	−4.0				
Diabase	200−630	0.68	0.2	630−800	0.9	0.8	*
"				800−1200	0.35	−1.2	
Norite-gabbro				640−1100	1.0	0.2	*
Eclogite	200−640	0.76	−1.25	140−970	0.66	−2.34	[182]

*Data obtained by É. I. Parkhomenko.

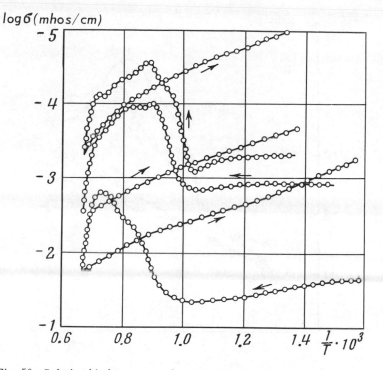

log σ (mhos/cm)

Fig. 50. Relationship between conductivity and temperature for three samples of serpentinite on heating and cooling.

The behavior of conductivity with temperature in rocks with
polymorphic changes in state may be more complicated. A minimum in conductivity at temperatures of 600 to 700°C has been observed for a few samples of serpentinite (see Fig. 50) [186]. The
behavior as temperature is reduced, as may be seen on Fig. 50,
provides a monotonically decreasing conductivity, σ. Noritomi
explains this anomalous behavior by a transformation of serpentinite into olivene at a temperature slightly above 600°C. The anomalous behavior might also be explained by the change in valence of
iron ions, or oxidation, or by the effect of the hydroxyl radical,
OH.

Considering that such an anomalous behavior is not observed
when the variation of conductivity with temperature for samples of
serpentine are used, Noritomi concludes that the high-temperature
behavior of olivene is explained by its structure. The basis for
this hypothesis is, first, the similar values for activation energy

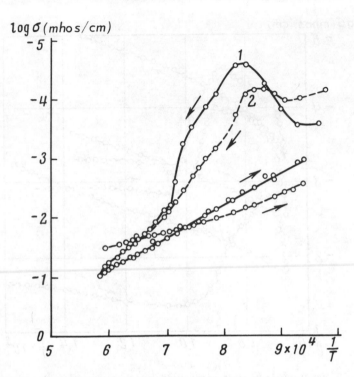

Fig. 51. Relationship between conductivity and temperature during cyclic heating and cooling: (1) Oshima lava, (2) Shirataki obsidian.

in serpentinite and olivene and, second, the similar abrupt increase in conductivity for both rocks at a temperature near 1100°C; that is, the character of the relationship $\log \sigma = f(1/T)$ reflects the importance of the chemical composition.

Murase [188] has observed anomalous behavior in the variation of conductivity with temperatures during the first temperature cycle for basalt, a series of lavas, and obsidian. In these rocks, as in the case of serpentinite, after the samples have been heated and subsequently cooled, the observed data fall reasonably well along a straight line (see Fig. 51), so that the relationship $\log \sigma = f(1/T)$ may be described by equation (III.10). The physical basis for this anomalous behavior is not apparent. Considering the similarity in behavior in serpentinite and in some of the lavas and obsidian, one might doubt the validity of Noritomi's explanation of the anomalous behavior of resistivity as a function of temperature.

Results of measurements of resistivity in dunite, olivene, granite, and pyroxenite to a temperature of 1200°C have been reported in reference [189]. In all of these rocks, resistivity was found to decrease by 6 or 8 orders of magnitude to nearly common values at 1200°C. Because the data were given as averages, it is not possible to evaluate possible changes in the conduction mechanism with increasing temperature. Moreover, the measurement procedure was not described.

The mechanism of conduction in rocks remains unexplained. Coster [182] has considered the particular characteristics of various types of charge carriers. In this reference, Coster applies Faraday's laws as a criterion for identifying charge carriers. The present author has come to the conclusion that electrons, as well as ions, play a role in conduction. Noritomi et al. [184–186], in analyzing their own data and earlier work, have come to the conclusion that in olivene or rocks with olivene structure, conductivity at temperatures below 600°C can be attributed to impurities, in the temperature range from 600° to 1100°C, conduction is by ionic or semiconduction processes, and at temperatures above 1100°C, conduction is ionic in nature. Noritomi [186] has compared the chemical composition of acidic and intermediate rocks with values for activation energy, and has drawn some conclusions concerning the relative roles of the oxides SiO_2 and Al_2O_3 in conduction. Murase [188] has also noted some degree of correlation between energy of activation and silica content.

Zakirova [191], based on the results of studies of more than 200 samples of various rocks and minerals, has suggested that at temperatures of 700 to 1150°C, that is, between the first and second changes in slope on the log $\sigma = f(1/T)$ curves, sodium and potassium ions serve as charge carriers. Moreover, in rocks with similar potassium content, she noted a correlation between electrical conductivity along the second segment and age. The greater the age of the rocks, the lower was its conductivity. This behavior, in the opinion of the investigator, is associated with the change in valence caused by the radioactive decay of K^{40} to Ca^{40}. Substitution of divalent Ca in the crystal lattice changes the physical properties of the lattice in such a way as to reduce the electrical conductivity.

The data which have been described show that temperature has a very pronounced effect on the resistivity of all rocks. In changing the ambient temperature from room temperature to 1000°C, resistivity changes by several orders of magnitude. Resistivities are widely different at low temperatures, but show less range at high temperatures.

The nearly similar and relatively low values for activation energy for rocks in the temperature range to 600 to 700°C indicate that conduction is by impurities, while at higher temperatures, conduction may be by semiconduction processes or by ions, as has been noted in references [189] and [191], which do not exclude the importance of the role played by Na and K ions.

Effect of Pressure on the Resistivity of Rocks

Studies of variations of the natural electromagnetic field of the earth indicate that the electrical conductivity of rocks changes with depth. Beginning at depths of the order of several tens of kilometers, the resistivity decreases with depth, with the rate of change being greatest at depths of 400 to 500 km, where the resistivity drops to a few Ω-cm (193, 194). It is important to know whether this variation in resistivity depends on chemical changes in composition of rocks in the upper mantle, or whether it is caused by the effects of temperature and pressure. Therefore, in order to study the properties of rocks in order to evaluate conditions deep in the earth, it is necessary to simulate the thermodynamic conditions which prevail at such depths. The reproduction of these conditions in labroatory studies is of value not only in the solution of problems in crystal studies, but are also of interest in exploration geophysics and well logging. At present, in view of the increasing depths to which wells are being drilled, it is necessary to study the effect of pressure on the resistivity of sedimentary rocks to aid in the quantitative interpretation of electric logs.

Investigations of the effect of pressure on resistivity of rocks have taken two directions. In one approach, the concern is that in well logging, the viewpoint is primarily that of finding the relationship between pressure and resistivity in reservoir rocks, with porosity a parametric factor. In these studies, pressure is limited to $p = 1000 \, kg/cm^2$. In the other approach (crystal studies) it is necessary to know the variation of resistivity with pressure for pressures up to several tens or hundreds of thousands of atmospheres. The development of theories about the conditions deep within the earth has been hindered by the lack of such data.

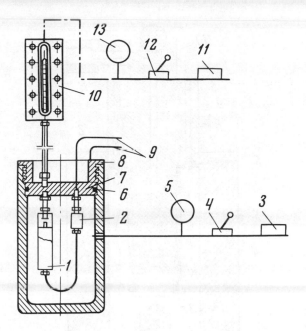

Fig. 52. Equipment for measuring the formation factor of
rocks under pressure: (1) sample in a Lucite sleeve, (2)
lucite, (3) reservoir, (4) hydraulic pump, (5) manometer
for measuring the increase in pressure, (6) ring, (7) seal,
(8) seal seat, (9) conductor to Wheatstone bridge, (10) in-
put, (11) reservoir, (12) hydraulic compressor, (13) manometer for measuring internal fluid pressure.

The relationship $\rho = f(p)$ in reservoir rocks has been studied
both by Soviet scientists [195-202] and by foreign scientists [203-
205]. The first of such studies was that done by Fatt [203]. In
particular, in these studies and in studies conducted by later stu-
dents, both the external pressure on the rock framework and the
internal pressure on the pore fluid were taken into consideration.

Because of this consideration, two pressure systems are
used in measuring the resistivity of detrital rocks at high pressures,
one to provide pressure to the rock framework and the other to
provide pressure to the pore fluid, with two manometers being used
to measure these two pressures. A sketch of the equipment used by
Fatt is shown in Fig. 52. Granville [205] has used similar equipment, which permits measurements to be made with any combination of axial and end pressures, or with uniform triaxially applied

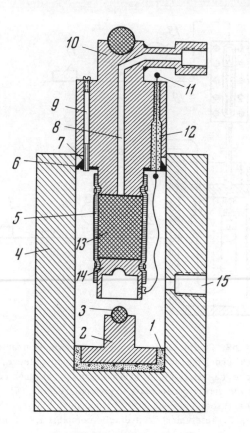

Fig. 53. High-pressure chamber: (1) textolite
ring, (2) base, (3) base sphere, (4) shell, (5)
"rubashka" resin, (6, 7) plastic ring and seat
for seal, (8) channel for entry and removal of
liquids from the sample, (9) channel for air to
leave, (10) piston, (11) conductors, (12) insulation, (13) sample, (14) lower electrode,
(15) channel for pressurizing oil.

pressure (Fig. 53), and pressures may be varied with or without
interaction. The rock is placed in a transverse press used to
load the end pressure. Oil is used as a medium with which to load
the rock with a hydrostatic pressure. This pressure is applied by
a pump through a port in the pressure vessel. In studies involving
interaction (internal pressures), the piston is connected to a second pump through pressure tubing.

In these studies, the samples must be insulated from the pressurizing medium, and this is done by using Lucite wafers and a plastic sleeve around the sample. A Wheatstone bridge is used to measure the resistivity at a frequency of 1000 cps.

Results from such studies, considered in terms of formation factors (the ratio of bulk resistivity to water resistivity), indicate that resistivity increases with increasing pressure on the rock framework.

The first resistivity-pressure studies done by Fatt [203] involved 20 standstone samples saturated with a sodium chloride brine. In all cases, the formation factor and, thus, the resistivity were found to increase with pressure. The relative sizes of the change in formation factor varied from sample to sample, being as much as 35% for an excess frame pressure of 350 atm.

The basic results obtained by Fatt were later confirmed by Wyble [204]. In addition, Wyble considered the effect of sample anisotropy and found it to be negligible. Granville [205], using the same pressures, measured formation factors not only for sandstones but also for limestones. He found an increase for the formation factors of sandstones and limestones of 125 and 132%, respectively, for a difference between frame and fluid pressures of 350 atm. In these early papers, there are too few data to allow consideration of the effect of cementation, but the amount of cement must play a significant role. A series of studies [195-199] have provided convincing data concerning this factor.

The experimental data, which are shown in Fig. 54 as the relation between relative resistivity, ρ/ρ_0 (the ratio of rock resistivity at a particular pressure to the resistivity of the same rock when no external pressure is applied), and pressure difference

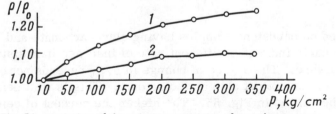

Fig. 54. Variation of the relative resistivity of a sandstone sample for a pressure of 50 atm on the saturating liquid: (1) shaly carbonate cement, (2) carbonate cement.

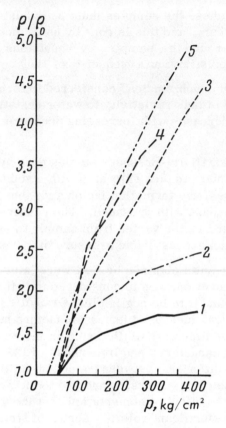

Fig. 55. Variation of the relative resistivity caused by pressure for samples with
shaly cement: (1) shaly sandstone, cement content 12%, (2) shaly sandstone,
cement 20%, (3) shaly sandstone, cement
28%, (4) shaly sandstone, cement 25%,
(5) shaly sandstone, cement 35%.

measured on sandstone samples having clay-carbonate and clay cement, clearly indicate a different rate of increase in resistivity
with pressure. The degree of change in ρ/ρ_0 depends not only on
the type of cement but also on the amount. This is well demonstra-
ted by the curves in Fig. 55. The higher the percent of cement,
the greater is the rate of increase in resistivity with increasing
pressure. The greater percentage change in relative resistivity is
usually observed over the pressure range from 1 to 200 atm.

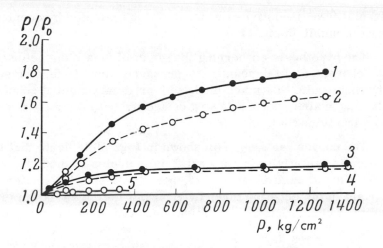

Fig. 56. Experimental curves for the relative variation in resistivity, ρ/ρ_0, with effective pressure (p): (1) Medina sandstone, 100% saturation, (2) the same with a residual saturation of 32%, (3) Torpedo sandstone, 100% saturation, (4) the same with a residual saturation of 15%, (5) synthetic alumina sample with 100% saturation.

Studying the effect of differences in the amount and type of cement in sandstones, Marmorshtein found increases in ρ/ρ_0 varying from 9 to 650% for pressures to 600 atm. The largest increases were found in rocks with significant amounts of clay cement. The effect of cement may be used to formulate the following series, reflecting the effect on the resistivity-pressure curve: by composition – carbonate, micaceous, hydromicaceous, and kaolinitic; by type – laminar, continuous cement, or basal. In addition, the author concludes that in some cases, the presence of cement will increase the resistivity (clay-carbonate cement) and in other cases, the presence of cement will decrease the resistivity (clay cement).

Other investigators [200, 201] have found increases in the resistivity of sandstones and carbonate rocks of 35 to 80% for pressures not exceeding 400 kg/cm^2.

Dobrynin [200] has postulated that small-diameter pores, associated with the presence of clay in a rock, play a basic role in minimizing the degree of increase in resistivity with pressure. The importance of the role of small-diameter pores has been confirmed by several studies [203, 205]. These studies have shown the porosity varies to a considerably smaller degree than do either permeability or resistivity, which are much affected by opening or closing of small-diameter pores.

The hypothesis concerning the effect of small-diameter pores on the character of the resistivity-pressure curve is further supported by the results of studies made with artifical samples consisting of alumina grains, porosity being controlled with the amount of dispersed fraction [200].

The curves for $\rho/\rho_0 = f(p)$ shown in Fig. 56 indicate that the variation in resistivity in such artificial samples is much less than that for natural sandstones. Also, pores with small diameter are important in the equation given by Dobrynin for the variation in resistivity, which relates the relative formation factor to pressure

$$\frac{P_p^p}{P_p} = \cfrac{1}{\left(2\cfrac{1 - \beta_p^{max} F(p)}{1 - k_p \beta_p^{max} F(p)} - 1\right) k^{f(p)\frac{C_{sh}}{k_p + C_{sh}}}},$$

where C_{sh} is the percent by volume of clay minerals in a rock, k_p is the porosity, and β_p^{max} is a coefficient for the maximum compressibility of pores in a rock.

Computations show that the second multiplying term in the denominator, which reflects the effect of clay content, plays a dominant role in determining the value for the relative formation factor. This is in good agreement with results obtained by Marmorshtein, which also show the importance of the clay content in a rock. However, it is necessary to consider both the amount of cement and the type as classified by Marmorshtein in his studies.

The relationship between the resistivity of clastic rocks and pressure exhibits hysteresis, particularly at low porosities, associated with permanent deformation in a number of cases [201].

Making generalizations from these data, we conclude that there is an increase in resistivity with pressure in sedimentary rocks saturated with water. The amount of increase depends on the amount and type of cement. The rate of increase in resistivity in such rocks with pressure depends on how small the pores are and on the shape of the pores, that is, on the shape of the current-carrying paths in which the permeability is greatly reduced when pressure is applied (50% for p = 350 atm).

These studies have shown that corrections for the effect of pressure must be applied when formation factors determined for rocks in place are compared with formation factors measured under laboratory conditions.

It should be noted that one important parameter – temperature – was not considered in any of these studies. Elevated temperatures would change the absolute magnitude of rock resistivity and, possibly, enhance the relationship between resistivity and pressure. The study of resistivity in sedimentary rocks with pressure applied and with temperature raised simultaneously would be of considerable practical importance.

The second approach to pressure studies – study of the electrical properties of igneous rocks at great depths and high pressures – has assumed considerable importance in the Soviet Union, but the first work in this area was a study by Hughes [206] of the relationship between the resistivity of peridotite and temperature and pressure.

At present, we are collecting data on the change in resistivity of various igneous rocks with uniaxial pressures to 600 kg/cm^2, hydrostatic pressures to 1000 kg/cm^2, and quasi-hydrostatic pressures to 40,000 kg/cm^2 being applied.

With uniaxial pressure, the behavior of the function $\rho = f(p)$ may be studied with pressures up to half the breaking strength of the sample. In such studies, the samples must have layers of parallel grains so that the pressure field is distributed uniformly through the rock.

The use of hydrostatic pressure requires the use of specialized high-pressure equipment. In the equipment used by Volarovich and Bondarenko [207], the pressure chamber consists of a heavy-walled hollow cylinder having an inside diameter of 40 mm, and two plugs (see Fig. 57). A sample is attached to the electrodes with a bitumen cement, and the electrodes are carefully insulated from the rest of the pressure chamber with Plexiglas.

The electrical contacts are connected through a shielded cable to the electrical measuring system. The plugs and samples are placed in the thick-walled cylinder and compressed with the working piston. Hydrostatic pressure is developed with a gas compressor, and nitrogen, which is used as a pressurizing medium, is

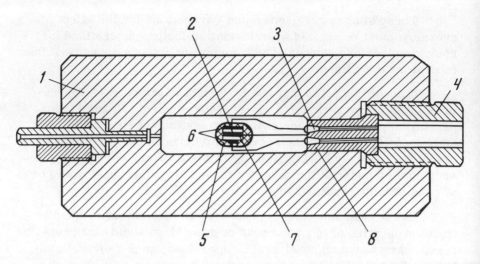

Fig. 57. Device for studing the resistivity of rocks under hydrostatic pressure to 1000 kg/cm^2: (1) shell of chamber, (2) sample, (3) conductors, (4) probe assembly, (5) rosin coating, (6) electrodes, (7) guard ring, (8) plug with conductors.

injected into the pressure chamber with a special injection system. The equipment is installed in a console in the horizontal plane.

In studying resistivity with quasi-hydrostatic pressure, equipment with an external heater, such as that shown in Fig. 58, is used

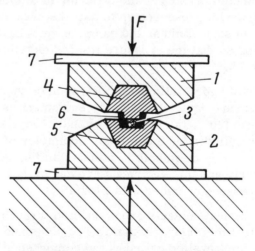

Fig. 58. High-pressure device: (1, 3) retaining blocks, (3) sample, (4) mortar, (5) pestle, (6) insulation, (7) mica.

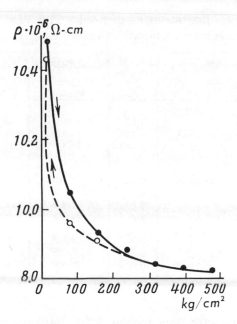

Fig. 59. Relationship between resistivity and uniaxial pressure in basalt during application and removal of pressure.

[208]. This equipment has the following components: a punch, a pressure block in which the sample is mounted, and two supporting members. The upper part of the system serves as one electrode, the lower part as the other. The punch is insulated from the pressure block with a ring of Plexiglas or ebonite (or mica for operating temperatures from 300 to 700°C), attached to the piston. The rock sample is also placed in a Plexiglas or ebonite ring with a thickness of 0.6 mm. At high temperatures, a heat-resistant ceramic ring is used. Thus, the pressure from the punch is applied both to the sample and to the Plexiglas, so that a three-dimensional pressure field, which is approximately hydrostatic, is developed. In making measurements at elevated temperatures, the whole system is placed in a furnace. The large amount of metal in the system assures a uniform distribution of heat in the sample.

These experimental studies have shown that there is a decrease in resistivity, both for uniaxial and hydrostatic pressures, for dense sedimentary rocks with low water content and for igneous rocks at pressures up to 1000 kg/cm^2.

Table 27. Resistivities of Rocks with Various Water
Contents with Uniaxial Pressure (p_{uni}) Applied

Rock	Water content by weight, %	ρ, Ω-cm		$\Delta\rho$, %
		for $p_{uni} = 4$ kg/cm^2	for p_{max}	
Sandstone $p_{max} = 507$ kg/cm^2	0.22	$8.7 \cdot 10^6$	$7.43 \cdot 10^6$	14.5
	0.17	$1.18 \cdot 10^9$	$0.61 \cdot 10^9$	48.8
	0.08	$1.32 \cdot 10^{11}$	$1.11 \cdot 10^{11}$	15.8
Basalt $p_{max} = 239$ kg/cm^2	0.62	$2.76 \cdot 10^7$	$2.5 \cdot 10^7$	9.5
	0.32	$7.79 \cdot 10^7$	$6.72 \cdot 10^7$	13.7
	0.16	$1.48 \cdot 10^9$	$1.33 \cdot 10^9$	10.2
	0.03	$7.84 \cdot 10^9$	$7.05 \cdot 10^9$	10
	0	$9.19 \cdot 10^9$	$8.15 \cdot 10^9$	11.2
Peridotite $p_{max} = 462$ kg/cm^2	0.1	$3.07 \cdot 10^5$	$2.93 \cdot 10^5$	4.5
	0.003	$4.07 \cdot 10^5$	$3.77 \cdot 10^5$	6.5
	0	$6.54 \cdot 10^5$	$6.04 \cdot 10^5$	7.6

We, along with Bondarenko [125], have completed a series of
measurements of the resistivity of various rocks after applying
pressures ranging from 10 to 600 kg/cm^2, using the direct current
method with a guard ring. A decrease in resistivity with increasing
pressure was noted in all cases, but not always to the same extent.
In some cases the change amounted to 10 to 20%, while in other
cases, it was only a few percent. In most cases, with uniaxial
pressure, most of the change in resistivity was found for pressures
below 300 kg/cm^2 (see Fig. 59). At higher pressures, the resistivity did not change significantly. Most of the physical properties
for a rock vary in the same manner with pressure.

The results of measurements made on sandstone, basalt, and
peridotite are listed in Table 27. These data, along with data obtained for other rocks, indicate that the greatest decrease in resistivity when pressure is applied is found for some intermediate water
content in a rock, rather than for maximum or minimum amounts
of water.

In particular, different rocks exhibit a maximum rate of
change in resistivity for different water contents, which is only natural because of the difference in shape and tortuosity of pore structures from one rock to another.

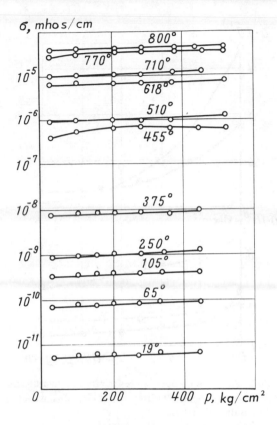

Fig. 60. Relationship between conductivity and pressure in basalt at various temperatures.

Our studies have indicated that temperature does not have a significant effect on the shape of the resistivity-pressure curve for uniaxial pressures. A series of isothermal curves – curves for the change in electrical conductivity as a function of pressure measured at a constant temperature – for diabase are shown in Fig. 60. An increase in conductivity with pressure is observed at all temperatures. For diabase, the curves $\sigma = f(p)$ lie nearly parallel to one another, with the change in resistivity caused by pressure being about 10% in most cases. In the case of basalt, there is some flatening of the isothermal curve, $\sigma = f(p)$, with increasing temperature. Over the temperature range 19 to 215°C, pressure changes resistivity by 40%, while at the higher temperatures, the change is only 20%

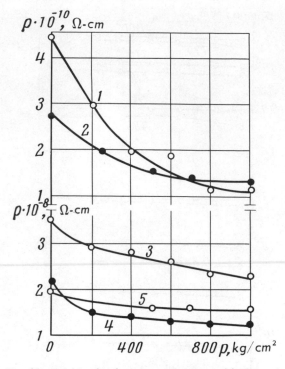

Fig. 61. Relationship between resistivity and hydrostatic
pressure for dry rocks: (1) limy sandstone, (2) sandstone,
(3, 4) basalt, (5) diabase.

or less. The tendency for the dependence of resistivity on pressure
to decrease at high temperatures may be partly explained by a decrease in polarization at high temperatures.

Considering these studies, it is apparent that the effect of
pressure on resistivity is much less than the effect of temperature.
Uniaxial pressures change neither the manner nor the degree of
change in resistivity with temperature. Curves for the function
$\log \sigma = f(1/T)$ for basalt at atmospheric pressure and under a hydrostatic pressure of 600 kg/cm^2 lie exactly parallel to one another.

Studies of resistivity in rocks subjected to hydrostatic pressures up to 1000 kg/cm^2 indicate that the relationship $\rho = f(p)$ determined with uniaxial pressures are essentially correct [207].
For measurements made with hydrostatic pressure, not only the

apparent but also the true resistance of a sample is determined using the formula

$$R_{true} = \frac{v - P}{I},$$

where P is the polarization voltage, computed from experimental
data.

Experimental data indicating the nature of the change in resistivity in rocks subjected to hydrostatic pressure are shown in
Fig. 61. A pronounced decrease in ρ takes place as the pressure
is increased to 200-400 kg/cm^2. The percentage change in resistivity evoked by hydrostatic pressure is larger than that caused by
the application of uniaxial pressure. In the one case, $\Delta\rho$ amounts
to 20-40% while in the other, it is only 5-20%. This obviously is
explained by the greater amount of pore closure and improved contact between grains which is obtained with hydrostatic pressure.
Values for true resistivity at two pressures are listed in Table 28.
With increasing hydrostatic pressure, the true resistivity varies in
exactly the same way as does the effective resistivity.

Data for the effect of hydrostatic pressure above 1000 kg/cm^2
is limited to that reported by Hughes [206], who found an increase in
the resistivity of peridotite, rather than a decrease as reported in
references [207] and [208]. Hughes carried out his experiments in
a pressure chamber with an internal furnace. The length of the
electrical furnace was 9 cm. Nitrogen was used as a pressurizing
agent. Samples in the form of cylinders, 0.17 cm in diameter and
0.3 cm high, were used. Measurements were made with alternating
current at a frequency of 400 cps and a voltage of 50 V. The dependence of resistivity on pressure was observed at three temperatures – 1063, 1143, and 1210°C. At each of these temperatures, the
resistivity was measured, for pressure increased and decreased,
through the values 1000, 2500, 4000, 5500, 7000 and 8500 kg/cm^2.
The averages of the values measured at each pressure, plotted in
Fig. 62, show a decrease in conductivity with increasing pressure.
The variation in resistivity with pressure was found to be uniformly
about 2.3-3.7% per 1000 kg/cm^2, not changing with temperature.
The total decrease in conductivity for pressures up to 8500 kg/cm^2
was only 30%, an amount of change which could have been caused by
a reduction in temperature by 14°C. Hughes' work indicates the

Table 28. Resistivity of Rocks Subjected to a Hydrostatic Pressure of 1000 kg/cm^2

Rock	True resistivity ρ, Ω-cm		$\Delta\rho$, %, at 1000 kg/cm^2
	at atmospheric pressure	at 1000 kg/cm^2	
Sandy marl	$11 \cdot 10^7$	$7.8 \cdot 10^7$	30
Limy sandstone	$2.45 \cdot 10^{10}$	$0.80 \cdot 10^{10}$	67
Sandstone	$1.9 \cdot 10^{10}$	$0.97 \cdot 10^{10}$	48
Slate	$0.89 \cdot 10^8$	$0.60 \cdot 10^8$	33
Basalt	$2.5 \cdot 10^8$	$1.96 \cdot 10^8$	22
Diabase	$0.13 \cdot 10^{10}$	$0.10 \cdot 10^{10}$	23
Peridotite	$2.4 \cdot 10^7$	$1.4 \cdot 10^7$	42

relative importance of the roles played by temperature and pressure in the variation of resistivity with depth in the earth. It is unfortunate that the effect of the pressurizing medium on the properties of the rocks has not been evaluated in past studies. It is known that the penetration of gas into a sample can significantly change the character of the experimental results [209].

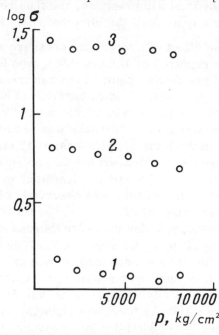

Fig. 62. Relationship between conductivity and pressure in peridotite for various temperatures: (1) 1063°; (2) 1143°, (3) 1210°C.

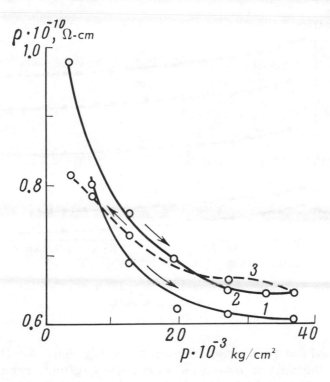

Fig. 63. Relationship between resistivity and pressure in diabase: (1, 2) second and third cycles of pressuring, (3) reverse path.

Studies of the resistivity of rocks to pressures of 40,000 kg/cm² conducted by us in collaboration with Bondarenko [208] led to the recognition of two different forms of the relationship between resistivity and pressure. For some rocks, there is a continuous decrease in resistivity with increasing pressure; in other rocks, the resistivity decreases at low applied pressures, but increases at higher pressures. The equipment used in these studies is that shown in Fig. 58, which provides quasi-hydrostatic pressure to a sample.

A decrease in resistivity was observed for all pressures up to 40,000 kg/cm² for samples of basalt, diabase, and amphibolite. In the case of diabase, a significant hysteresis was observed, with very good agreement being observed between values measured on

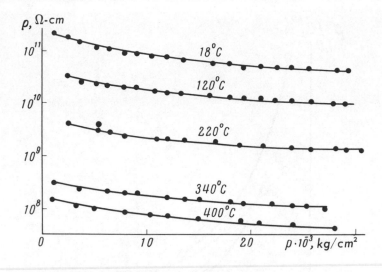

Fig. 64. Relationship between resistivity and pressure for basalt at
various temperatures.

the second and third pressure cycles (see Fig. 63). Resistivity
changes the most at pressures below 20,000 kg/cm², remaining
nearly constant at higher pressures. The resistivity of one diabase
sample changes by about 30% in going from a pressure of 3000
kg/cm² to one of 38,000 kg/cm². In the case of another diabase
sample, the change was 2.5 times as great, about 78%.

The resistivity of basalt was observed to decrease more uniformly than that of diabase. The change in resistivity of basalt
samples was 60 to 80% for an increase in pressure from 3000 to
33,000 kg/cm². Isothermal curves obtained with such basalt samples at a pressure of 24,000 kg/cm² are very nearly parallel to
each other (Fig. 64). This indicates that in the temperature range
from 18 to 400°C, temperature changes do not affect the relationship
$\rho = f(p)$.

Plotting these experimental results in a different manner (log
σ and $1/T$) permits separation of the roles played by these parameters individually. As may be seen in Fig. 65, the relationship
between conductivity and temperature for basalt is that given in
equation (III.10). The activation energy computed for the isobaric
curve for p = 2600 kg/cm² is 0.52 eV. Two other isobaric curves

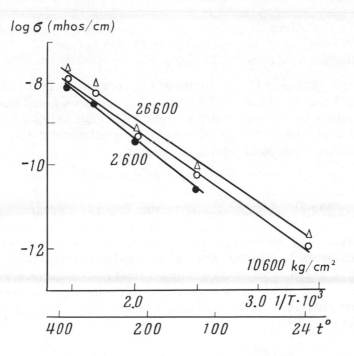

Fig. 65. Relationship between resistivity and temperature for basalt at various pressures.

are parallel to each other; thus the activation energies computed
at pressures of 10,600 and 26,600 kg/cm² are both 0.436 eV.
Thus, there is a slight tendency for activation energy to decrease
with increasing pressure, the decrease amounting to 16% in this
case. This suggests that it would be worthwhile to make a care-
ful study of the resistivity-temperature relation as a function of
pressure. The low values for activation energy indicate that, at
the temperatures considered, conduction is by impurity ions.

Considering the data shown in Figs. 64 and 65, it is apparent that an increase in temperature to 300-400°C causes a much
larger effect on σ than an increase in pressure to tens of thousands
of atmospheres. Increasing the temperature to 400°C causes an increase in resistivity of several orders of magnitude, while an increase in pressure from 2600 to 26,000 kg/cm² causes no more
than a 70% change in resistivity. In the author's most recent
studies, he has established that there is a decrease of 60 to 80%

in the resistivity of granite, gabbro, peridotite, serpentinite, and pyroxenite, over the range 1300 to 21,000 kg/cm², at a temperature of 600°C.

A decrease in resistivity at high pressures is not observed for all rocks, as was noted earlier. It has been found that a number of rocks exhibit a more complicated relationship for $\rho = f(p)$. Such rocks include augite porphyry, serpentinized dunite, and pyroxenite, from one investigated region.

In these rocks, resistivity decreases to a minimum value and then increases as pressure is increased. The minimum resistivity in one sample of augite porphyry was found at a pressure of 17,000 kg/cm². In two other samples of the same rock, the same sort of behavior was observed, except that the minimum occurred at pressures below 12,000 kg/cm². As pressure is reduced, usually the pressure-resistivity curve shows a linear segment (see Fig. 66). In a number of cases, the minimum in resistivity was observed at lower pressures for measurements made during the reduction of pressure.

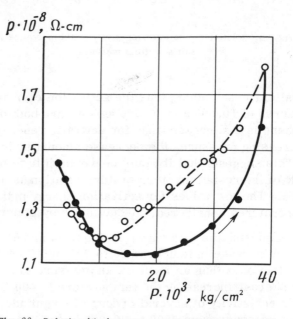

Fig. 66. Relationship between resistivity and pressure for augite porphyry during application and removal of pressure.

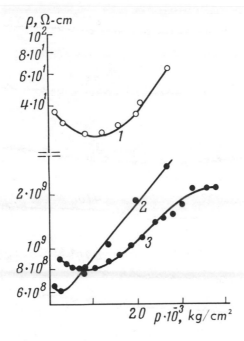

Fig. 67. Relationship between resistivity and pressure for serpentinized dunite at various temperatures: (1) 24°, (2) 356°, with sustained pressure, (3) 356°C without pressure being sustained.

A similar behavior was observed with serpentinized dunite. With this rock, some drift in effective resistivity with time was observed for measurements made with direct current. Therefore, measurements were made with continuously rapidly increasing pressure, with temperature being raised in steps. In other measurements, in which no drift in resistivity was apparent, the pressure was maintained constant while the temperature was varied. The curves obtained for resistivity as a function of pressure should be the same for both techniques. Differences were found in the numerical values of resistivity measured in the two manners (Fig. 67). In experiments with increasing temperatures; the general character of the relation $\rho = f(p)$ is invariant.

A similar effect — a slight upward shift in resistivity at constant pressure, both at room temperature and at elevated temperatures —

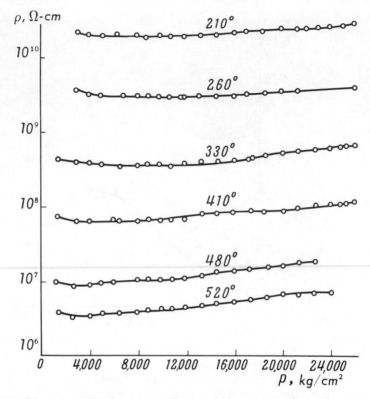

Fig. 68. Relationship between resistivity and pressure for the mineral
microcline cut parallel to the cleavage at various temperatures.

has been found by Kozyrev [210] in studies with monocrystalline
and polycrystalline selenium. This effect is apparently caused by
a decrease in polarization at the instant heat is applied, with a
subsequent increase under constant pressure.

A minimum resistivity is observed in minerals as well as in
rocks. In studies with the minerals nepheline and microcline, I
have found a similar minimum in resistivity over some range in
pressure. Thus, the general nature of the relationship between
resistivity and pressure in microcline does not vary over the temperature interval 210–520°C, as shown in Fig. 68. The activation
energy calculated from two isobaric curves for 10,600 and 21,300
kg/cm² are nearly the same, about 0.9 eV. Therefore, the observed
increase in resistivity is not caused by a change in activation en-

ergy, and must be caused by a decrease in the mobility of the current-carrying particles.

In the case of nepheline, the pressure for which a minimum is observed decreases as temperature is increased, and at temperatures of 620-680°C disappears completely. The resistivity is nearly constant for pressures up to 4000 kg/cm^2, and increases uniformly at higher pressures.

Petrographic examination of polished rock sections, prepared after experiments were completed, indicates that there is no change in structure caused by the application of pressure [208]. In addition to fragmentation and deformation of grains of some minerals, twinning in pyroxene grains and formation of deformed borders in serpentinized dunite are found.

It may be assumed that these changes in the structure of a rock have no important effect on the resistivity-pressure curve, because the resistivity returns to its initial value after a temperature cycle in many cases. The maximum difference in the values following the first and second cycles is no more than 50%.

The existence of a minimum resistivity for some pressures has been observed not only by Bondarenko and me, but also by Moiseenko, Istomin, and Ushakova [211] with measurements of resistivity for basic and ultrabasic rocks (basalt, serpentinite, olivinite, peridotite, pyroxenite, and dunite) for uniaxial pressures to 20,000 kg/cm^2.and hydrostatic pressures to 2000-3000 kg/cm^2. In some rocks, this minimum occurs over a narrow range of pressures; in other rocks, the minimum resistivity is fairly constant over a broad range in pressure. There is some reason to doubt the validity of minima observed during the initial heating phase, because such a minimum may be caused by changes in contact resistance between the electrodes and the rock.

In other studies conducted by Moiseenko and Instomin [212] at temperatures of 440 and 500°C, only a decrease in resistivity with increasing pressure was observed for olivinite, serpentinite, and eclogite. The change in resistivity for serpentinite on increasing the pressure to 30,000 kg/cm^2 at a temperature of 440°C was more than three orders of magnetude. Such large decreases in ρ as a function of pressure requires correction. In an important paper by Bradley, Jamil, and Munro [213], there also was reported a decrease in resistivity in fayalite and spinel for an applied pressure of 35,000 kg/cm^2 at a temperature of 680°C.

In studies made with water-bearing granite, the resistivity is found to decrease greatly with increasing pressure [190] in the high pressure range. The resistivity of granite was studied over a pressure range from 1 to 9000 kg/cm^2 and a temperature range from 600 to 1200°C.

It has been established that the effect of the partial pressure of the pore water on resistivity, beginning with 4000 kg/cm^2, is markedly suppressed. If at a temperature of 900°C, the resistivity varies by five orders of magnitude for a pressure change from 1 to 4000 kg/cm^2, then the resistivity changes only slightly for a pressure change from 2000 to 8000 kg/cm^2 (see Fig. 68).

Because granite contains only a small amount of water (less than 1%) under natural conditions, the data for the resistivity of water-bearing granite over the specified temperature and pressure ranges may be used in interpreting geophysical data only in special cases.

The effect of pressure on the conductivity of dielectrics has been studied very little, either theoretically or experimentally. Therefore, in order to obtain the data which would be necessary to develop a theory for the composition and physical properties of materials deep in the earth, it is necessary to carry out a diverse program of research on the electrical properties of the fundamental rock-forming minerals over a wide range of pressures and temperatures. Evaluation of the various relationships obtained for different rocks and minerals would permit an explanation of the physical basis for the phenomena described above. Consideration should also be given to the effect that the presence and type of ore mineral might have. Most ore minerals are semiconductors; they may have a significant effect on the shape of the resistivity-pressure curve for a rock if present in significant amounts.

As has been noted eariler, semiconductors are characterized by a different relationship between resistivity and pressure [25-29, 192]. In some semiconductors, resistivity is increased by the application of pressure, while in others, resistivity decreases with pressure. The difference in behavior of various semiconductors is explained by changes in activation energy and in the mobility of charge carriers.

By analogy with the behavior of semiconductors, one might suppose that the effect of pressure on the resistivity of dielectric materials would be to increase it in others. This might explain the different forms of the resistivity-pressure curves which are observed.

The material which has been summarized above is of primary interest in geophysics because it clarifies the relative importance of the roles played by temperature and by pressure in changing the resistivity of rocks deep in the earth. The pressures which have been used in our researches (40,000 kg/cm^2) correspond to depths of the order of 150 km; they change the resistivity by only 70%. An increase in temperature to 400°C (corresponding to a depth of 10 or 12 km) causes a decrease in resistivity of a number of rocks by several orders of magnitude. Thus, the pressures which have been used in laboratory studies correspond to far greater depths than do temperatures of 400°C.

From data reported in references [181-186], it is known that resistivity may decrease by 5 or 6 orders of magnitude when the temperatures is raised to 800°C. Considering the overriding effect of temperature on the resistivity of rocks, future work should be directed to the study of pressure effects at very high temperatures, such that the thermodynamic conditions deep in the earth are duplicated in the laboratory.

The following generalizations may be made based on results of investigations of the effect of high pressures on the resistivity of rocks. A marked increase in resistivity is observed for sedimentary rocks with a high water content when pressures to 600 kg/cm^2 are applied. This is caused by a decrease in the cross-section of current-carrying paths and by closure of pores. The degree of change in resistivity depends on the porosity of the rock and the composition of the cementing material.

The resistivity of dense sedimentary rocks with low water contents and of igneous rocks decreases when a uniaxial pressure is applied. The most rapid change in resistivity with pressure is found at low pressures, in the range of 200-400 kg/cm^2.

Two different forms of the relation between resistivity and pressure are observed when samples are subjected to large, quasi-hydrostatic pressures. In some rocks, resistivity decreases

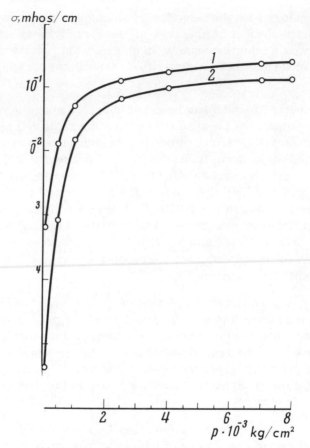

Fig. 69. Relationship between conductivity and partial pressure on the pore water for granite [190]: (1) at a temperature of 1100°C, (2) at 900°C.

uniformly while pressure is raised to 40,000 kg/cm^2; in other rocks, the resistivity first decreases, then increases as the pressure is raised. In order to explain the physical basis for these observed relations, it is necessary to collect more experimental data, both for rocks and for minerals, for comparison with data on the chemical composition and the mineralogical characteristics of rocks.

Chapter IV

Dielectric Loss in Rocks

Brief Introduction to the Theory of Dielectric Loss

In alternating electric fields, a dielectric material is frequently characterized by its dielectric loss, which depends on the fraction of electrical energy lost to heat, rather than by its electrical conductivity. The energy loss occurs as the result of two processes: conduction and slow polarization currents. Because dielectric materials always possess some degree of conductivity, dielectric loss is observed both in direct current flow and in alternating fields. The distinction is that for direct current flow the loss depends only on the conductivity, but in alternating fields, loss may take place with displacement currents also.

The effect of displacement currents on dielectric loss depends on the relaxation time for the polarization and the frequency range. Polarization by electron or ion displacement takes place in a very short time – 10^{-14}-10^{-15} sec or 10^{-12}-10^{-13} sec, respectively. Thus, loss from this type of displacement current does not occur at radio frequencies. The time required for relaxation polarization to take place is considerably longer.

Slow development of relaxation-type polarization gives rise to currents which decrease with time. This current may be considered to consist of two parts – resistive and reactive:

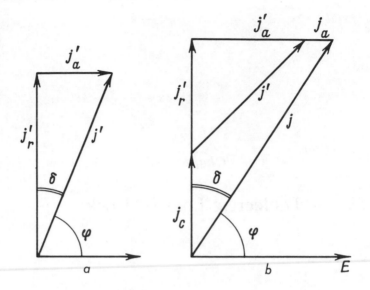

Fig. 70. Vector diagrams for current density and electric field intensity in a dielectric: a — with the presence of only relaxation polarization; b — relaxation polarization and electron displacement.

$$j' = j'_a + j'_r,\tag{IV.1}$$

with the first component being in phase with the applied voltage, and the second being out of phase through an angle $\pi/2$ (see Fig. 70a). The development of heat in a dielectric by dielectric loss is caused by the resistive current component and is independent of the reactive current.

The resultant current, which is the vector sum of the two orthogonal components j_a' and j_r', is out of phase with the applied electric field through a phase angle, φ. The angle δ, which is the complement of the phase angle φ, is termed the dielectric loss angle.

The tangent of this angle is the ratio of resistive current to reactive current

$$\tan\delta = \frac{j_a'}{j_r'}.$$

In actual dielectric materials, equation (IV.1) is complicated by the presence of out-of-phase conduction and displacement currents. In such cases, the total current in a dielectric consists of three terms: a capacitance current (displacement current) j_c, a current j', which has a resistive and reactive component, j'_a and j'_r:

$$j = j_c + j_a + j''_a + j''_r.$$

Graphically, the total current may be represented in vector form as shown in Figure 70b.

The dielectric loss tangent for an actual dielectric material is

$$\tan \delta = \frac{j_a + j'_a}{j_r + j_c}.$$

The current which contributes to loss is usually taken into account by introducing a complex dielectric constant [13, 217, 218]

$$\varepsilon^* = \varepsilon' - i\varepsilon''. \tag{IV.2}$$

With direct current ($\omega = 0$), the dielectric constant has only a real part, assuming the extreme value, ε_0. At an infinitely high frequency, the dielectric constant is completely real also, but assumes a minimum value, ε_∞.

In a general case with alternating current at a frequency ω:

$$\varepsilon^* = \varepsilon_\infty + \frac{\varepsilon_0 - \varepsilon_\infty}{1 + i\omega\tau}, \tag{IV.3}$$

where τ is a relaxation time.

Separating this expression into real and imaginary parts, we have

$$\varepsilon' = \varepsilon_\infty + \frac{\varepsilon_0 - \varepsilon_\infty}{1 + \omega^2\tau^2},$$

$$\varepsilon'' = \frac{(\varepsilon_0 - \varepsilon_\infty)\,\omega\tau}{1 + \omega^2\tau^2},$$

$$\tan \delta = \frac{(\varepsilon_0 - \varepsilon_\infty)\,\omega\tau}{\varepsilon_0 + \varepsilon_\infty\omega^2\tau^2}. \tag{IV.4}$$

This relationship is called the Debye formula. The relationship $\varepsilon'' = f(\varepsilon')$ may be plotted in the complex plane for use in analysis and extrapolation of experimental data.

In Figure 71, the magnitude of the real quantity ε_∞ is plotted as a line segment OA along the abscissa, and the real magnitude $\varepsilon_0 - \varepsilon_\infty$ is plotted as the line segment AC. Points which satisfy equation (IV.3) are located symmetrically on a circle with the center at the point $(\varepsilon_0 + \varepsilon_\infty)/2$. The tangent of the angle ϑ is equal to $\omega\tau$. The loss term ε'' assumes a maximum value at a frequency $\omega = 1/\tau$, when dipole polarization drops to half its maximum value. The length of the line segment OB is equal to the complex dielectric constant. Such a diagram permits determination of relaxation time and the range of relaxation effects in the dielectric constant.

Dielectric loss, which causes heating in a dielectric material, may be characterized in terms of the angle δ and $\varepsilon'' = \varepsilon \tan \delta$, as well as in terms of specific loss; that is, the energy expended per cubic centimeter, which may be computed with the formula

$$P = \frac{\varepsilon'' f}{1.8 \cdot 10^{12}} E^2,$$

where f is the frequency and E is the applied electric field.

The higher the quality of a dielectric material, the lower will be the values for $\tan \delta$, δ, ε'', and P.

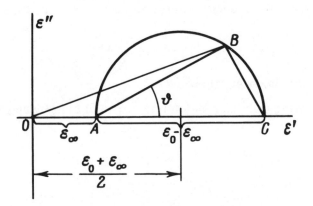

Fig. 71. Argand diagram for a complex dielectric constant.

Relationship of Dielectric Loss and Dielectric

Constant to Frequency

In many areas of science and technology, including the area of geophysics, knowledge of the relationship between dielectric constant, loss tangent, frequency, and temperature is important. The nature of the dependence on frequency is primarily determined whether loss is caused by conduction currents or slowly occurring polarization processes. Also, in heterogeneous dielectrics, the mutual dispositions of the components comprising the material are important.

For dielectrics with high conductivity and weak polarization processes, tan δ is inversely proportional to frequency, satisfying the equation

$$\tan \delta = \frac{4\pi \cdot \sigma}{\omega \varepsilon_{\infty}} = \frac{1.8 \cdot 10^{12}}{\varepsilon_{\infty} f \rho} \, ,$$

where $f = \omega/2\pi$ is the frequency in cps, and ρ is the resistivity in Ω-cm. In this case, the dielectric loss and resistivity are practically equal to the zero-frequency values and do not vary with frequency. The dielectric constant also remains over a broad range of frequencies.

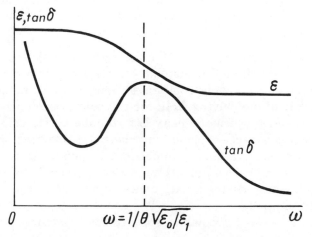

Fig. 72. Dependency of ε and tan δ on frequency for dielectrics with high resistivity and dielectric loses.

In cases in which there is little conductivity in dielectric material, and in which a relaxation process is well developed, the behavior of tan δ as a function of frequency is as follows: at low frequencies, conduction dominates over relaxation processes, and tan δ decreases with increasing frequency at very low frequencies. As frequency is increased further, relaxation processes become relatively more important and tan δ increases until the period of the driving field becomes equal to the relaxation time for the process (Fig. 72). At that point, the resistive current component j' assumes its maximum size because the current does not drop during a half-period. The reactive current component increases but not linearly with frequency, and, as a result, the dielectric constant decreases slowly with frequency [12].

Considering that tan δ is the ratio of resistivity to reactive currents, their different rates of change with frequency lead to the development of a maximum in tan δ. Further increase in frequency causes a decrease in tan δ because the reactive current increases in linear proportion to frequency when ε =constant (see Fig. 72).

Dielectric loss in a material with a well-developed relaxation process is described by the Debye formula (IV.4).

A pronounced increase in tan δ at optical frequencies, caused by resonance, may be observed. Resonance loss takes place when the driving field frequency equals that natural frequency for electrons. The tan δ-frequency curve exhibits a maximum which does not shift with changes in temperature.

The tan δ-frequency curve for dielectric loss due to relaxation phenomena has the following characteristics. At low frequencies, the value of ε remains constant so long as there is time during a half-cycle of the driving field for the polarization process to be completed. At frequencies near $1/\tau$, the dielectric constant exhibits dispersion because a phase shift between polarization and the driving field, which is frequency dependent, is found. At very high frequencies, when slow polarization processes do not have time to develop significantly during a half-cycle, the value of ε is constant, assuming its minimum value ε_∞. The character of the change in dielectric constant with frequency for various types of polarization was shown in Fig. 1.

In cases in which a dielectric is inhomogeneous (a simple example would be a laminated capacitor), accumulations of charge may be found at the surface between unlike materials. The dielectric constant for such a material varies with frequency in a manner similar to that for simple orientation polarization described by the Debye theory. The loss factor ε'' for a two-layer composite dielectric is ·

$$\varepsilon'' = \varepsilon_\infty \left(\frac{\tau}{\omega \tau_1 \tau_2} + \frac{k\omega\tau}{1 + \omega^2 \tau^2} \right),$$

where $k = (\varepsilon_0 - \varepsilon_\infty)/\varepsilon_\infty$, and τ_1 and τ_2 are the relaxation times for the materials comprising the two layers in the composite dielectric.

The dielectric properties of a material consisting of two layers, each of which is divided into a large number of thin layers which alternate with each other but add up to the same total thickness as in the preceding case, will be exactly the same as the dielectric properties of the simpler two-layer material. This is also true for the resistivity of such composite materials. However, a change in the form of orientation of the components with respect to the direction of the driving field leads to change in $\tan \delta$ and ε''. The relationships between $\tan \delta$ and frequency for various forms and orientations of particles in a medium are shown in Fig. 73.

Thus, it appears that the texture of a rock may have a significant effect on the dielectric loss tangent.

Relationship Between Dielectric Loss, Dielectric

Constant, and Temperature

Dielectric loss and dielectric constant are affected significantly by changes in temperature. The manner of change depends on the nature of the loss mechanism. If relaxation processes are present only weakly, the relationship between loss tangent and temperature is exponential because conduction loss plays the primary role in such materials. Taking $\sigma_t = \sigma'_0 e^{-E_0/kT}$, we have

$$\tan \delta = \frac{4\pi\sigma_0}{\omega \varepsilon_\infty} e^{-\frac{E_0}{kT}},$$

where E_0 is the activation energy.

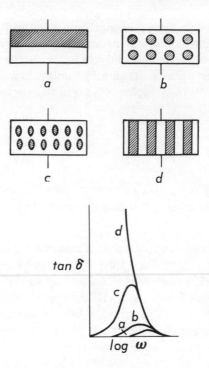

Fig. 73. Relationship between tan δ
and frequency for dielectrics with various forms and orientations of elements
(from Sellars): a — a two-layer dielectric
consisting of a conducting layer and
insulating layer; b — conducting spheres;
c — conducting ellipsoids; d — conducting
cylinders.

If the free conduction in a material is slight and relaxation
polarization is present, a maximum loss tangent as a function of
temperature is observed at that temperature at which the relaxation time is close to a half-period of the applied electric field.
For temperatures which cause the relaxation time to be greater
than or less than the half-period of the applied field, the loss is
less. The temperature dependence for tan δ in such cases is
shown in Fig. 74.

Free conduction may suppress the maximum in tan δ significantly, but with increasing frequency its effect diminishes and the
maximum in tan δ is shifted toward a higher frequency because

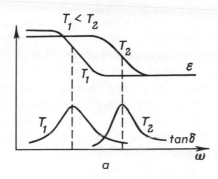

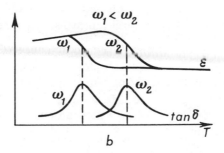

Fig. 74. Frequency and temperature dependence of ε and tan δ caused by relaxation processes in dielectrics: a — frequency dependence for two temperatures,
T_1 and T_2; b — temperature dependence
for two frequencies, ω_1 and ω_2.

relaxation time decreases with increasing temperature. Knowing
the temperature T_1 and T_2 on the Kelvin scale and the corresponding frequencies f_1 and f_2 at which maximum loss is observed, one
may compute the activation energy using the following formula:

$$E_0 = \frac{2.3\left(\log f_2 - \log f_1\right)}{T_2 - T_1} kT_1 T_2,$$

where k is Boltzmann's constant, which is $1.38 \cdot 10^{-16}$ ergs/deg,
$0.863 \cdot 10^{-4}$ eV/deg or 1.98 cal/mole-deg, depending on the units
desired for E_0. Then, the expression $\nu = \pi f e^{E_0/kT}$ is used to determine the natural vibration frequency for the particles taking
part in the relaxation process.

Variation of dielectric constant with temperature may not occur if relaxation processes are weak and conduction is large. In the inverse case, according to theory, a mobile maximum may be found which can be explained phenomenologically as follows. An increase in temperature is known to cause a decrease in relaxation time. Thus, the higher the temperature, the more fully will the polarization process be completed. This causes an increase in ε. At high temperatures, the relaxation time becomes less than a half-period of the applied field and the rate of change in ε starts to decrease with increasing temperature.

If no relaxation processes are present, the dielectric constant of a material is only slightly dependent on temperature, and this dependence is caused by a change in the number of molecules per unit volume with temperature, that is, a change in density.

In describing the temperature dependence of dielectric constant, the temperature coefficient $[1/\varepsilon \; (d\varepsilon/dt), \text{per degree}]$ is used. It gives the relative change in the dielectric constant for a one degree change in temperature. In gases, the value is found using the diffusion equation $\varepsilon = 1 + 4\pi n\alpha$ at temperatures, where α is the electron polarization.

Differentiating with respect to t, we have:

$$\frac{1}{\varepsilon}\frac{d\varepsilon}{dt} = -\frac{\varepsilon - 1}{t}.$$

The dependence of ε in a neutral liquid on temperature is similar to that for a gas, being caused by a decrease in the number of molecules per unit volume. Differentiating the Causius—Mosotti relation (II.1) with respect to temperature, we have

$$\frac{1}{\varepsilon}\frac{d\varepsilon}{dt} = -\frac{(\varepsilon - 1)(\varepsilon + 2)}{3\varepsilon}\beta_0,$$

where β_0 is the coefficient of volume expansion.

The temperature dependence of the dielectric constant in polar liquids is characterized by a maximum at a temperature at which the viscosity of the liquid is sharply reduced, so that the molecules may be easily oriented. Further increase in temperature increases the thermal agitation of the molecules, so that they do not remain oriented as long. In such cases, the temperature coefficient $[1/\varepsilon$

$(d\varepsilon/dt)]$ is found by graphical differentiation of the curve $\varepsilon = f(t)$, measured at a single frequency [111].

In solid dielectrics, this coefficient may be either positive or negative, depending on whether the dominating effect is decreasing density or increasing polarizability. In most cases, ionic crystals are characterized by positive values. Examples are KCl, NaCl, and corundum.

Solid dielectrics (cellulose, waxes, polymers, and so on), which exhibit relaxation polarization, are characterized by very strong temperature dependences for the dielectric constant. It should be noted that the frequency of the applied field has a pronounced effect on the ε-temperature curve.

Equipment and Methods for Determining Dielectric Constant and Dielectric Loss

Various techniques have been using in measuring the dielectric constant and loss with samples having large dimensions: bridge methods, resonance methods, and heterodyne methods. The use of a particular method depends on the range of frequencies, the properties of the material, and the required precision [219].

In determining the dielectric properties of a material, the proper choice of electrodes is as important as the proper choice of measuring equipment. The requirements for electrodes for use in AC measurements are more severe than requirements for electrodes for DC resistance measurements. Electrodes must have a low resistance because the use of electrodes with a resistance of several hundred ohms, as in the case of graphite electrodes, causes an increase of as much as 0.001 for the loss tangent, to be computed when the electrode resistance and sample capacitance are added in series. Also, the electrodes must adhere closely to the sample. A small layer of air between the electrodes and the sample will contribute a capacity in series with that of the sample. This leads to a decrease in the total capacity and tan δ. At high temperatures, the nature of the electrodes may have the opposite effect. Some investigators [220, 221] have noted that silver electrodes increase the dielectric loss at high temperatures. One cause, obviously, would be diffusion of silver into the sample. The best results, according to Panchenko [82], are obtained with gold electrodes sputtered on to the sample in a vacuum.

Along with these factors, the size of the sample and the detail construction of the measuring equipment are important in the precision of measurements.

The capacity of a sample increases with its cross-sectional area and decreases with its thickness, so that the capacity of a thin wafer may be measured most accurately. Commonly, samples are prepared in the form of thin discs, with their diameters many times greater than their thicknesses. In deciding on the sample cross-section, the size of the individual mineral grains must be considered, as well as the condition of quasi-stationarity [222]. This condition requires that the diameter of the capacitor containing the sample be less than the wavelength of the applied electric field. The radius of a quasi-stationary condenser must satisfy the condition

$$R \leqslant \frac{0.24\lambda}{2\pi \sqrt{\varepsilon}} = 3.8 \cdot 10^{-3} \frac{\lambda}{\sqrt{\varepsilon}},$$

where λ is a wavelength.

The simpliest form of sample holder is a planar capacitor in which the edges of the electrodes are at the same radius as the edge of the sample. In making precise measurements, a guard ring is used to prevent distortion of electric field lines around the edge of the sample. A diagram of such a sample holder is shown in Fig. 75.

If the sample thicknesss is of the same order as the sample diameter and the electrode diameter, it is necessary to compute the magnitude of the edge capacity. The edge capacity correlation for a capacitor with square parallel plates of width a may be computed with the formula [222a]

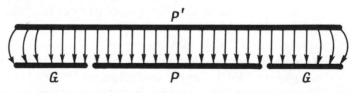

Fig. 75. Sketch of the construction of a simple parallel plate condensor with a guard ring: P and P' — discs, G — a ring, lying in the same plane as P, with P being in the measurement circuit.

$$C = \frac{a}{3.6\pi h}\left[1 + \frac{h}{\pi a}\left(1 + \ln 2\pi \frac{a}{h}\right)\right],$$

or in the case of circular electrodes, with the formula

$$C = \frac{R^2}{4h} + \frac{R}{4\pi}\left[\ln \frac{16\pi (h + b) R}{h^2} + \frac{b}{h} \ln \frac{h + b}{b} - 3\right],$$

where R is the electrode radius, b is the electrode thickness, and h is the spacing between them.

Bridge Methods. Various types of bridges, with either parallel or series resistances and capacitances, are widely used for measuring the dielectric constant, loss, and resistivity at audio frequencies. A bridge with parallel resistances and capacitances is normally used for measuring the properties of high-loss rocks.

The use of a bridge circuit with series or with parallel connection of resistance and capacitance depends on the equivalent circuit used to represent the dielectric, with ideal resistances and capacitances connected either in series or in parallel.

In studying high-loss materials, a bridge with parallel connections is used.

In studying dielectrics with low loss, a bridge with series connections is used.

In collaboration with Volarovich and Valeev [223], I have used the bridge shown in Fig. 76a for measurements of the electrical resistivity of sedimentary rocks. In this bridge, the impedences of each arm are

$$Z_0 = R_0, \quad Z_1 = \frac{R_1}{1 + i\omega C_1 R_1},$$

$$Z_2 = \frac{R_2}{1 + i\omega C_2 R_2}, \quad Z_4 = \frac{R_x}{1 + i\omega C_x R_x}.$$

The balance conditions require that

$$R_x = \frac{R_0 \cdot R_1}{R_2} \frac{1 + C_2^2 R_2^2 \omega^2}{1 + R_1 R_2 C_2 C_1 \omega^2}.$$

T. L. Chelidze [27, 224] has used the substitution form of the Schering bridge (Fig. 76b) for measuring the electrical properties

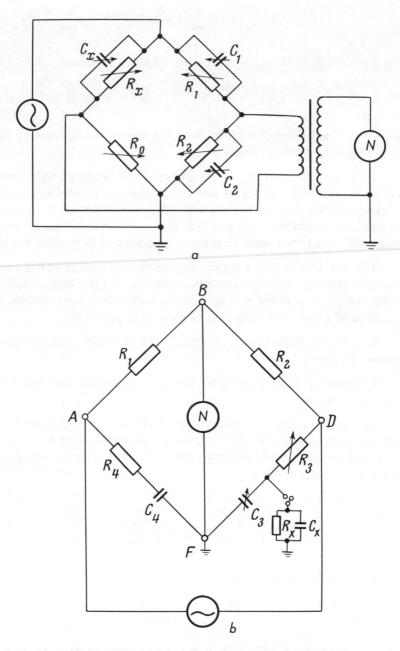

Fig. 76. Bridge circuit: a — for measuring resistance, b — universal, N — null
indicator.

of rocks with tan $\delta \leq 0.1$. This bridge requires the following sequence of observations. The resistance R_3 and capacity C_3 are
measured initially with the sample. Then, the sample is connected
into arm DF, unbalancing the bridge so that R_3 is rebalanced to
R_{31} and C_3 to C_{31}.

The resistance R_x and capacitance C_x of the sample are computed with the following formulas

$$R_x = (R_3 - R_{31}) + \frac{1}{(R_3 - R_{31}) C_3^2 \omega^2} \ ,$$

$$C_x = \frac{C_3}{(R_3 - R_{31})^2 C_3^2 \omega^2 + 1} - C_{31},$$

where ω is the angular frequency.

A resistance without a reactive component must be used in
measurements with alternating current. A vibrating galvanometer
or a standard null meter of type INO-3M may be used as a null
indicator in the frequency range 100-1000 cps, or an oscilloscope
may be used.

Detailed descriptions of a number of special bridge circuits
may be found in the literature [225-230].

The resonance method is used at frequencies from 10^5 to 10^8
cps. There are several resonance systems which differ in the
manner in which the resistive component is determined. The various methods are variable resistance, variable conductance, variable admittance, variable frequency, and observation of increase
in voltage at resonance. In all methods, micrometer electrodes
must be used and the lengths of connecting cables minimized. A
detailed description of these methods and measurement techniques
is given in the monograph [219]. The various measurement techniques and the frequency ranges over which they are applicable
are listed in Table 29.

Various commercially available instruments may be used in
measuring the AC electrical properties of rocks. The model E10-7
impedance bridge may be used in measuring reactance in the range
from 15 Ω to 100 kΩ with a precision of 2-5% and capacity in the
range from 20 to 100 pF with a precision of 0.5%. The model
UM-3 universal bridge has a wider range of applicability for measuring resistance (from 0.1 to 5×10^6 Ω) and capacity (from 10 pF

Table 29.   Frequency Ranges and Measurement Techniques [217]

Frequency, cps	Measurement techniques
Radio	
$30-10^4$	Bridge circuits
$10-10^7$	Schering bridge
10^4-10^8	Resonance method
10^8-10^{11}	Standing wave method
Infrared	
$10^{11}-10^{14}$	Interference method
Optical $- 10^{14}$	" "

to 100 μF). In addition, the loss tangent may be measured over
the range from 10^{-3} to 0.1. Measurements are made at frequencies of 100 and 1000 cps. The model E12-4 universal bridge,
"immortel," provides a higher degree of precision. It may be
used to measure capacitance, inductance, and conductance over
the following ranges: capacitance, 0.0002-11.1μF; conductance,
0.01-111.0 mΩ. The operating frequency is 1600 cps.

The Effect of Frequency and Temperature on Die-
lectric Loss and Dielectric Constant in Minerals

At present, studies of the dielectric properties of minerals
have been directed mainly toward the accumulation of data on the
absolute value of the dielectric constant. As a result, observations
of the temperature and frequency dependence of the dielectric loss
tangent and the dielectric constant have been limited. However, such
data are of great scientific and practical value. The scientific
interest is based on the fact that the form of the relationships ε and
tan δ = F (f, t), are less affected by the texture of a dielectric than
is the absolute value, ε_0. Knowledge of these relationships permits study of polarization mechanisms, the nature of dielectric
loss for a particular internal structure, and a number of other
physical parameters for a dielectric – activation energy for dipole
groups, angular frequency of rotation for relaxing dipoles, true
relaxation time, distribution of relaxation times, and so on.
At the same time, the increased use of AC methods in electrical
exploration requires an understanding of the frequency dependence for ε and tan δ in rocks.

As has been noted earlier, rocks are complicated aggregates
of many phases. Therefore, the practice of relating observed

physical properties and their behavior to the physical properties of the major component is not particularly fruitful. In analyzing the behavior of the bulk properties of a rock, it is important to have data on the properties of all the minerals comprising the rock.

Mica is the material which has been most intensively studied with respect to the effect of temperature and frequency on the dielectric constant and loss angle, but quartz, feldspar, halite, gypsum, talc, rutile, and other minerals have also been studied to a lesser extent. Among this group, there are only a few of the common rock-forming minerals in igneous rocks.

A detailed study of the properties of mica has been carried out by Vodop'yanov [231-236]. There are also papers by a number of other authors [237-241]. Vodop'yanov investigated various effects on the dielectric constant and loss in micas – muscovite and phlogopite – taken from different ores. These studies indicated that the properties of mica depend not only on the particular type of mica but also on the nature of the inclusions [232].

It has been established that the loss tangent in pure muscovite is small and does not depend on frequency. The presence of inclusions causes an increase in loss which diminishes with increasing frequency. The rate of decrease in tan δ is more rapid when the external field is directed parallel to the cleavage than when it is normal to the cleavage. Later, Vodop'yanov in collaboration with Vorozhtsova [236] investigated the temperature dependence of ε and tan δ as a function of the chemistry of the inclusions. The loss tangent in muscovite mica containing a large amount of water of crystallization has a well-developed maximum as the temperature is raised through 350°C, which is larger than the largest maximum caused by impurities in the mica (see Table 30).

A second heating of muscovite with limonite inclusions leads to a diminution of the maximum, or, in extreme cases, its complete suppression. With a third cycle of heating, the relationship

Table 30. Effect of Impurity Content on the Maximum

Impurity content, %	Maximum value, tan $\delta \cdot 10^4$
5.4	None
25.0	5.8
37.0	9.3
40.0	11.0

tan δ = F (t°) assumes the character which is typical of pure muscovite (Fig. 77). Vodop'yanov suggests that the disappearance of
the maximum is explained by the evolution of most of the water of
crystallization from the limonite, the water being a polar molecule
which can give rise to relaxation polarization.

The observation of a constant value for the dielectric constant
while the loss tangent exhibits a maximum as a function of temperature (Fig. 77) is very difficult to explain, because the physical

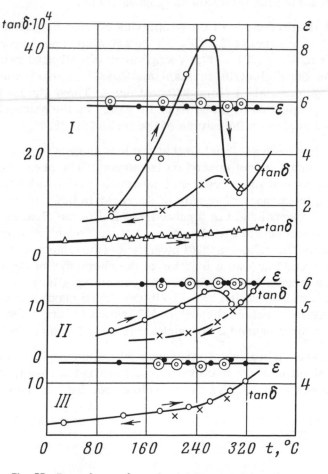

Fig. 77. Dependence of ε and tan δ on temperature for
muscovite with limonite inclusions at a frequency of 10^6
cps: (I, II, III) cyclic measurements; the line with triangles
is the temperature curve for a sample of clean muscovite.

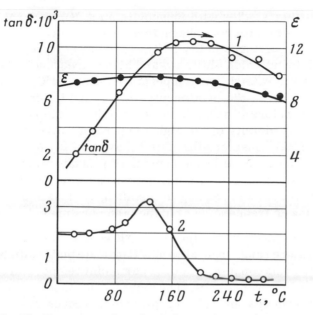

Fig. 78. Temperature dependence for ε and tan δ for muscovite
samples with biotite inclusions: (1) tan δ while heating the samples, (2) tan δ after heating at 400°C.

processes involved in developing a maximum in loss tangent with
a relaxation phenomenon require a simultaneous change in the dielectric constant. It is of interest to note that heating the muscovite
sample to 600°C causes a reappearance of a maximum. Vodop'yanov
and Vorozhtsova are of the opinion that this maximum is caused by
strongly bonded water of crystallization which is freed on heating.
This idea is in agreement with the fact that some water evolves on
heating muscovite to 300-400°C, but most of the water evolves at
a temperature 600-700°C. These data are applicable only for
muscovite with inclusions of limonite. The behavior of the relationship ε, tan δ = F (t°) is different for other types of inclusions. For
example, muscovite with inclusions of biotite exhibits a broad
maximum in tan δ, rather than a sharp maximum, and the dielectric constant increases slightly and then decreases with increasing
temperature (Fig. 78). Tempering such samples at a temperature
of 400°C changes the shape and amplitude of the temperature-induced maximum in the loss tangent. No maximum is observed with
samples tempered at 600° or 800°C, and the loss tangent increases
continuously as temperature is raised to 300°C, while the dielectric
constant remains constant.

In further studies with muscovite mica tempered at 600°C, two maxima with different characteristics have been noted [241, 242]. At temperatures below 94°C, a typical relaxation maximum in tan δ is found which shifts toward higher frequencies as the temperature is raised. At temperatures above 94°C, an anomalous behavior is found – the maximum shifts toward lower frequencies as the temperature is raised. Also, the dielectric constant decreases in the vicinity of the high-temperature maximum. Vorozhtsova [241] considers that this behavior of tan δ and ε may result from a decrease in the number of relaxators. According to the theory of Lyast, relaxation time increases as the density of relaxators decreases, leading to a shift in the maximum for tan δ toward lower frequencies at higher temperatures.

Observational data on the dielectric constant of muscovite over the temperature range from 20 to 520°C indicate that the dielectric constant remains constant only over particular temperature ranges, varying in different manners. The increase in ε is most pronounced at audio frequencies, and is less rapid at high frequencies.

In addition to muscovite, the dielectric properties of phlogopite mica have been investigated extensively [232, 234, 238, 239]. The basic difference between the chemical compositions of phlogopite and muscovite is the substitution of MgO for a large part of the Al_2O_3. The relation of the dielectric constant and loss to the frequency for phlogopite, taken from Vodop'yanov's data, is shown in Fig. 79.

As seen from these curves, there is a single maximum in the loss tangent at audio frequencies, accompanied by a decrease in the

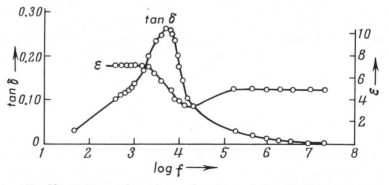

Fig. 79. Frequency dependence of ε and tan δ for phlogopite mica.

dielectric constant. In contrast, Keymeulen [239] has found two
frequency-dependent maxima. Two temperature-dependent maxima at low frequency have been reported in reference [237, 238].
The first is found at a temperature of -50°C, the second at 200°C.
The dielectric constant increases near the low temperature maximum in the loss tangent, and decreases at temperatures near the second maximum. Tempering phlogopite at temperatures of 500-700°C
does not change the magnitude or the position of the maxima in tan
δ. However, tempering at temperatures above 700°C leads to an
increase in the maximum of the loss tangent. Izergin has suggested
that this is caused by a change in structure. He explains the first
maximum by rotation of the polar group (OH) or of molecules of
water of crystallization, and the second by the transition from
water of crystallization to free water.

Opinions concerning the basis for temperature- and frequency-derived maxima in tan δ are conflicting in a number of cases.
According to Vodop'yanov, the existence of such maxima depends
mainly on the orientation of molecules of water of crystallization.
Keymeulen [239] attributes the presence of low-temperature maxima to the development of pseudodipoles by structural defects.

In investigations of the dielectric properties of gypsum and
talc [234, 235, 242, 243, 244], similar maxima in loss tangent
are observed as a function of temperature and frequency. At high
frequencies (λ =15 m), two maxima are found for the loss angle in

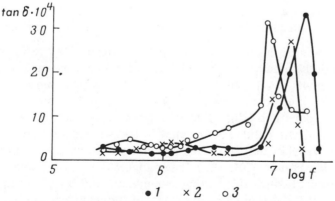

Fig. 80. Frequency dependence of tan δ for crystalline gypsum
at various temperatures: (1) 293°K,(2) 243°K,(3) 193°K.

gypsum, one at a temperature of 55°C and the other at 84°C [242, 243]. The presence of temperature-derived maxima at these same temperatures has been substantiated in another paper [244]. At a somewhat lower frequency (λ =300 m), the temperature-derived maximum is less pronounced. The behavior of the frequency-derived maximum in tan δ is typical of a material with relaxation polarization (Fig. 80). The relaxation process in these cases is the same as the process in mica, being explained, apparently, by the presence of water of water of crystallization. It is of interest to note that after tempering gypsum at 243°C, only one frequency-derived maximum occurs, that being for f=10^6 cps.

The existence of two maxima in the loss tangent has been explained by Vodop'yanov as being due to [243, 245] the two activation energies for water of crystallization. Similar maxima in tan δ are observed for talc [235]. The dielectric constants for gypsum and talc remain constant over the frequency range 10^5-10^7 cps for temperatures of 20 to 140°C, despite the maxima in tan δ .

Thus, the fundamental result contained in the investigations described above on the dielectric properties of mica, gypsum, and talc is that the presence of water of crystallization may have a profound affect on the amplitude of the maximum in both the temperature dependence of dielectric loss and in the frequency dependence.

The temperature-derived relaxation maximum in tan δ is explained either by orientation of hydroxyl radicals with various activation energies or by other processes in ceramic materials and salts [246-248]. The maximum in tan δ for salt is small and has been found to be invariant in a number of investigations [247].

Studies of the dielectric properties of the clay minerals kaolin and dickite have shown a different temperature dependence for tan δ [249, 250]. In the case of kaolin, a pronounced maximum in tan δ is found, which shifts toward higher frequencies at higher temperature (Fig. 81). When water is removed completely from kaolin, the relationship tan δ =f (t) is a straight line. This is the main proof that the source of the maximum is the presence of the polar radical OH. The loss tangent for dickite decreases almost linearly with increasing frequency, and as the temperature is increased from 140 to 332°C, it increases nearly tenfold.

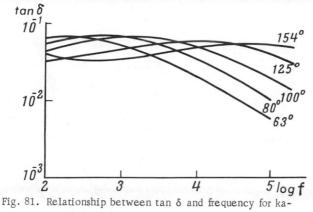

Fig. 81. Relationship between tan δ and frequency for kaolinite at various temperatures.

The dielectric properties of quartz along the several axes
have been studied as a function of temperature and frequency by
many investigators [251-254]. Bogoroditskii and Fridberg [251]
measured the value of ε at a temperature of 150°C. Stuarts [252]
has found a change in ε at some very high temperatures (to 380°C),
and in two later papers [252, 254], data for measurements of ε to
temperatures of 600-700°C were given. The results of these studies

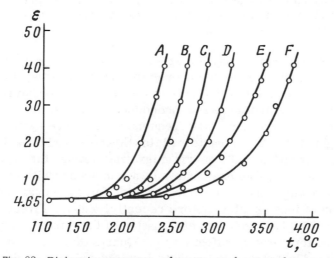

Fig. 82. Dielectric constant ε_3 of quartz as a function of temperature and frequency: A = 1, B = 4, C = 8, D = 20, E = 40, F = 90
kcps.

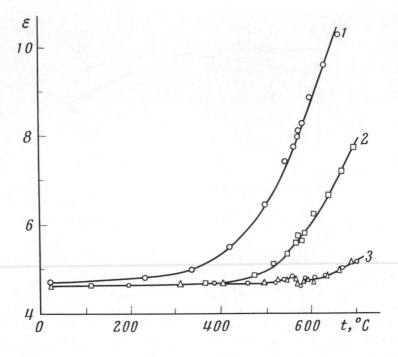

Fig. 83. Effect of the removal of impurity ions on the temperature dependence of ε_3 in quartz: (1) original sample, (2) after first cleansing, (3) after second and third cleansing.

are not entirely consistent. Stuarts has found that ε remains constant for temperatures to 400°C over the frequency range 1-90 kcps, while ε_3 has a constant value only at the lowest temperatures. The temperature at which there is a pronounced increase in ε_3 depends on frequency. With increasing frequency, the curve $\varepsilon_3 = F(t)$ shifts to the right (Fig. 82). Stuarts explains this observed behavior in terms of impurities in the quartz migrating through the crystal lattice. However, this explanation is not in accord with experimental data which show the same behavior for ε_3 in quartz before extraction of ions as after. More recently, Zubov, Firsova, and Molokova [253] have given an explanation of the effect of impurity ions on the temperature effect on ε_3. Quartz of ultimate purity was obtained by extracting impurity ions by applying an electric field of 1000-25,000 V/cm in cycles along the optical axis at a temperature of 600-700°C. As seen in Fig. 83, as impurity ions are extracted from quartz by successive cleansing cycles, the

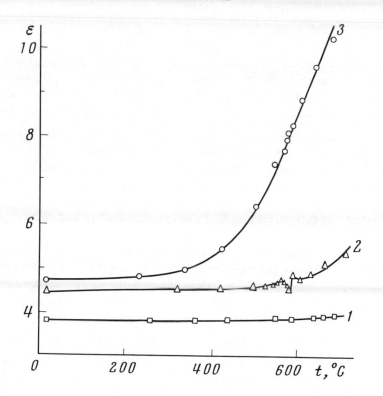

Fig. 84. Temperature dependence of the dielectric in crystalline and fused quartz: (1) fused quartz, (2, 3) single crystals, corresponding to ε_1 and ε_2.

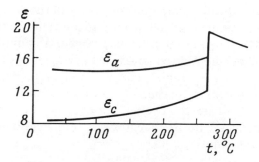

Fig. 85. Relationship between dielectric constant and pressure in monocrystalline boracite.

$\varepsilon = F(t)$ curve flattens, and after the third cycle, a peculiar variation in ε_3 associated with a phase transition may be seen. In the original crystal, it was masked by the effects of the impurities. Also, the nature of the relationship $\varepsilon_3 = F(t)$ becomes the same as the relationship for ε_1 obtained by the same investigators and shown in Fig. 84.

Thus, the increase in ε_3 at elevated temperatures reported by Stuarts [252] and Bondarenko [254] may be explained by the presence of impurities. Bondarenko observed a pronounced increase in the dielectric constant ε_1 for quartz in the temperature range between 570-600°C. The lack of correspondence between the nature of this variation of ε_1 with temperature and that reported in reference [253] apparently resulted from the use of different frequencies. Bondarenko made his measurements with a very low frequency – 5 kcps rather than the 1 Mcps used in the research of Zubov and colleagues [253]. Thus, in the one case with $f = 1$ Mcps, there is only a slight tendency for ε_1 to increase at temperatures of 600-700°C, and in the other case, it is pronounced.

The dielectric properties of molten quartz differ significantly from those of crystalline quartz. The dielectric constant for molten quartz doesn't change with either frequency or temperature over wide ranges of temperature and frequency [255]. The frequency-temperature behavior of tan δ indicates an inverse proportionality to frequency. This is a characteristic of ohmic loss. A maximum in relaxation processes is found in simple glass at low temperatures, 50-80°K. V.A. Ioffe [256] has advanced the hypothesis that the relaxing particles are elements of a silicate lattice.

Boracite exhibits anomalous behavior in the dielectric constant similar to that for crystalline quartz at the temperature of a phase transition [257] (see Fig. 85). Also, boracite exhibits a strong anisotropy in the dielectric constant. At a frequency of 500 cps, the dielectric constants ε_c and ε_a are 8.2 and 14.1, repectively. The dielectric constant increases with increasing temperature, but the spread between the two values decreases. The dielectric constant decreases with temperature at temperatures above the phase transition point.

V. A. Ioffe and his colleagues have arrived at a number of

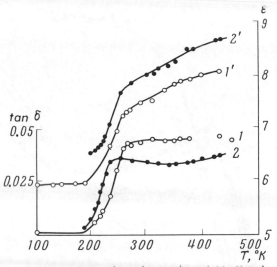

Fig. 86. Temperature dependence of tan δ (1, 2) and ε (1', 2') in orthoclase. (1, 1') 800,000 cps, (2, 2') 80,000 cps.

interesting conclusions based on studies of the dielectric properties of grains of various minerals in the aluminosilicate group
(albite, orthoclase, plagioclase, labradorite, microcline, and
cancrinite) [258-260].

All measurements were made in a vacuum after lengthy heating at 200°C, also in a vacuum. For most of the minerals, similar behavior was observed for dielectric loss and dielectric constant as a function of temperature: these parameters maintained
nearly constant values for temperatures up to 200°K, and in the
temperature range from 200 to 300°K, both the dielectric constant
and the loss increase rapidly (see Fig. 86). With further increase
in temperature, tan δ varies in different manners while ε continues
to increase. An exception is the temperature behavior for the loss
tangent in labradorite, for which there is a characteristically large
increase in tan δ at very high temperatures and also in the neighborhood of 400°K in the relationship ε = F (t) for one sample of cancrinite. In the case of cancrinite, ε was found to decrease with
increasing temperatures at frequencies of 500 kcps and 1.78 Mcps.
It should also be noted that the temperature dependence of tan δ for
the feldspar minerals at frequencies of 500, 1000, and 5000 cps is

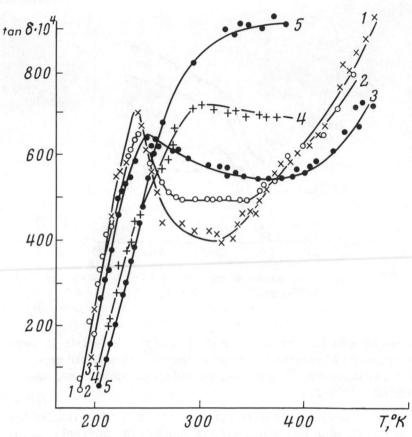

Fig. 87. Temperature dependence of tan δ in microcline cut parallel to cleavage. Frequency in cps: (1) 500, (2) 1000, (3) 5000, (4) 50,000, (5) 500,000.

such that there is a maximum which, as shown in Fig. 87, shifts toward higher temperatures at the higher frequencies and becomes flatter. Temperatures did not exceed 500°K in the experiments.

The frequency dependence of tan δ for all of the minerals studied was such that a maximum was observed at frequencies of 10^5–10^6 cps, the frequency not depending on temperature. This fact led the authors to suggest the presence of resonance polarization. The frequency dependence of tan δ for microcline, labradorite and cancrinite exhibits two resonance maxima (see Fig. 88). V. A. Ioffe and I. S. Yanchevskaya [in 258] have related the second

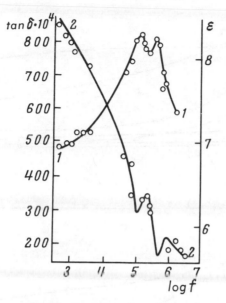

Fig. 88. Frequency dependence of
δ (1) and ε (2) for microcline cut
parallel to cleavage, measured at
room temperature.

resonance to the electrons in tetrahedral iron oxide. On the basis
of the maximum value of tan δ and its frequency dependence at low
frequencies, the authors advanced a working hypothesis that resonance arises as a result of a complicated interaction between
excess valence electrons in the aluminosilicate tetrahedra and
defects in the crystal lattice – cation vacancies or other disturb-
ances in electrical neutrality.

The dielectric constant in these aluminosilicates leads to
frequency dispersion. As frequency is increased from 10^3 to 10^7
cps, the dielectric constant for albite, measured either parallel to
or normal to the basal cleavage, and for plagioclase and labradorite
decreases by 25-80%. Also, the frequency dependence of ε for
cancrinite, microcline, and one sample of plagioclase (see, for
example, Fig. 88) exhibits a maximum which is obviously the result of resonance polarization. The mechanism for resonance is
a tentative hypothesis, in the opinion of the authors, and requires
further experimental study.

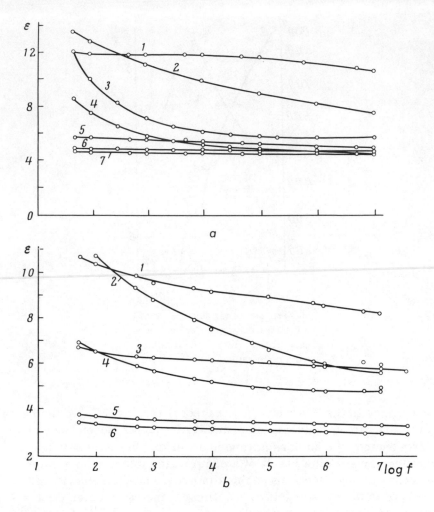

Fig. 89. Relationship between dielectric constant and frequency for various
minerals: a — (1) apatite, (2) dolomite, (3) chalcedony, (4) oolitic limestone,
(5) pegamatite quartz, (6) novaculite, (7) quartz cut perpendicular to optic
axis; b — (1) spodumene, (2) microcline, (3, 5) perthite, (4) talc, (6) orthoclase.

    The frequency dependence of the dielectric constant of a large
number of minerals has been studied by Howell and Licastro [14].
They measured the dielectric constant of 23 minerals over the frequency range from 100 cps to 10 Mcps at room temperature. As
a result of these studies, it was established that Brazilian quartz

($\parallel$ and $\perp$ to the optical axis), biotite, phlogopite, lipidomelane, muscovite ($\parallel$ to basal cleavage), calcite ($\parallel$ and $\perp$ to C axis), novaculite, gypsum ($\perp$ to b axis), orthoclase, and perthite exhibit little or no dispersion in the dielectric constant. In a number of other minerals, including dolomite, chalcedony, spodumene, microcline, and talc, a significant variation of dielectric constant with frequency was observed (Fig. 89). For most of the minerals, most of the variation in ε was obeserved at frequencies below 10^4-10^5 cps. At higher frequencies, there is very little change in dielectric constant.

The frequency dependence of the dielectric constant in apatite is anomalous. In contrast to the behavior in other minerals, at low frequencies (10^2-10^4cps) ε maintains a constant value, decreasing at higher frequencies.

There are some data available for the temperature dependence of ε and tan δ for synthetic rutile containing CaO and TiO_2. The dielectric constant for such rutile decreases from 1400 to 300 as the frequency is raised from 50 cps to 110 kcps, and increases greatly with increasing temperature [261, 262]. Over the temperature range for which the most change in dielectric constant is observed, the loss tangent goes through a maximum.

The dielectric properties of dielectric materials with a rutile lattice are determined by the quantities and chemical properties of the impurities. Small amounts of impurities cause distortions in the crystal lattice which cause relaxation processes involving weakly bonded ions [14, 261].

In conclusion, we may draw the following generalizations about the relationship of the dielectric properties of minerals to frequency and temperature.

For mica, talc, and gypsum containing water of crystallization, a maximum is found in tan δ as a function of frequency which is shifted toward higher frequencies as the temperature is raised, except in the case of muscovite mica. The existence of a frequency-derived maximum for tan δ in these minerals is related to the presence of water of crystallization. A similar maximum is found in the case of kaolin. Two temperature-dependent maxima in tan δ are observed for gypsum and talc, which are caused, apparently, by different activation energies for polar molecules.

Frequency-dependent maxima for tan δ caused by resonance polarization are observed in the aluminosilicate group of minerals. The loss tangent for these minerals shows a characteristic rapid increase at temperatures of 200–300°K, with the exception of labradorite, in which the change in tan δ takes place at higher temperatures.

The dielectric constant for most of the minerals that have been studied varies most rapidly at low frequencies. The dielectric constant for quartz and novaculite remains constant over a wide range of frequencies, while for phlogopite, the low frequency variation in ε is accompanied by a maximum in tan δ.

With increasing temperature, the dielectric constant for various minerals behaves differently. In some minerals, as, for example, talc and gypsum, it remains constant at relatively high temperatures, while in others, it begins to increase rapidly at different temperatures. The temperature dependence of ε in minerals is a function of the frequency of the applied field and of the presence of impurities.

For minerals in the rutile and perovskite groups, where typically the dielectric constant is very large, there is a temperature derived maximum in tan δ, and ε increases with increasing temperatures but decreases with frequency.

Investigations with minerals in the feldspar group and quartz have shown that the dielectric properties measured in different crystallographic directions are not the same, and do not vary with temperature and frequency in similar ways.

Table 31 lists values of loss tangent measured on various minerals at similar frequencies and temperatures for comparison.

Relationship of Dielectric Constant, Dielectric Loss, and Resistivity of Rocks to Frequency

The application of electrical prospecting methods, particularly induction and radio wave methods, presently requires knowledge of the frequency dependence of the electrical resistivity and dielectric constant of various rocks. Moreover, investigation of these relationships has great value in explaining mechanisms of polarization— contributing to the dielectric constant and providing a basis for a theory for the behavior of rocks in AC fields.

Table 31. Values of Loss Tangent for Minerals

Minerals	Temperature, °C	Frequency, 10^{-5} cps	tan δ	Reference
Muscovite	22	10	$0.3 \cdot 10^{-3}$	[111]
Phlogopite	22	10	$1.5 \cdot 10^{-2}$	[111]
Talc	—	10	$8.0 \cdot 10^{-4}$	[235]
Pyrophyllite	—	10	$1.0 \cdot 10^{-3}$	[219]
Gypsum	20	10	$2.0 \cdot 10^{-4}$	[235]
		10^2	$6.0 \cdot 10^{-4}$	[234]
Albite	27	5	0.13	[259]
Plagioclase	27	5	0.04	[260]
Microcline	27	5	0.076	[260]
Orthoclase	27	8	0.04	[260]
Cancrinite	27	5	0.03	[260]
Kaolin	63	1	$0.7 \cdot 10^{-2}$	[250]

A number of papers have been published describing studies of the electrical properties of rocks as a function of frequency [223, 224, 263-270], but the data reported by different investigators are often contradictory.

In the earliest studies, the object of study was sand or soil. The researchers established that dispersion was present in these and in other cases.

Smith-Rose [263] first noted the effect of water content on the character of frequency dependence for the resistivity and dielectric constant for soils. He obtained a large degree of dispersion for the values of ρ for minor moisture contents, with the amount of dispersion decreasing with increasing water content; and for dielectric constant, he found the opposite relationship: dispersion increased with increasing moisture content. In a later paper, Chakravarty and Khastigir [264] found only a weak frequency dependence for ε and ρ in soils, while Balygin and Vorob'ev [265] in measuring the dielectric constant and resistivity in soils generally found no dispersion. The disagreement between results obtained by various researchers may relate first of all to varying amounts of moisture in the samples and to differences in measurement techniques.

The first work which was a comprehensive study of the electrical properties of rocks in AC fields was that of Tarkhov [18].

Experimental data on the electrical properties of various rocks with different water contents measured at audio frequencies were given. This paper covered the basic considerations concerning the relationship of the dielectric constant and electrical resistivity to frequency.

From data obtained with an experimental setup especially designed to avoid contact effects, that author concluded that there is no dispersion in ε or ρ for rocks at audio frequencies. In Tarkhov's opinion, the surface capacitance associated with the collection of ions at the electrode surfaces and the decrease of this capacitance with frequency could explain the dispersion in ε and ρ found by some investigators [18].

Electrochemical processes at the contact between a metal and an electrolyte and the collection of charge in the portion of a dielectric close to the electrode contacts, giving rise to low-frequency polarization, are well known physical effects and must be excluded in determining dielectric constant and loss in rocks. On the other hand, the absence of frequency-dependent dispersion in ε and ρ at audio frequencies, based only on Tarkhov's work, cannot be considered to be positively assured. Because of the lack of a positive answer, various investigators have studied and are still studying this problem [14, 254, 165, and others].

I will review data from the literature on the frequency dependence of the properties of rocks below in three groups: sedimentary, metamorphic, and igneous.

Sedimentary Rocks. The nature of dispersion in electrical properties for sedimentary rocks depends primarily on the water content and on frequency. Also, processes taking place at the contacts between metal electrodes and water in the rocks and other surface effects may have a pronounced effects on the form of the relationships ρ, $\varepsilon = F(f)$. It must be realized that in many of the studies of dispersion in the electrical properties of rocks, the amount of water present was not controlled or the electrode effects were not considered. Only a few carefully made studies in which all these factors were considered are known, providing information on the true character of the frequency dependence for ρ and ε. Ivanov [268], in studying dispersion in the resistivity of sedimentary rocks, has used a method based on detecting the relative amounts of heat developed in a sample by current flow at various frequencies. Such an electrothermal method permits determination of the resistivity corresponding to conduction current flow. Such a system excludes all surface effects which may cause errors in the character of the frequency dependence for the rock resistivity. However, Ivanov did not consider that the conduction current actually is the sum of two components when slow polarization takes place – the true conduction component and the conduction component of the relaxation current caused by slowly occurring polarization. This second component may be the result of surface effects. The conductive component of the relaxation current, according to the data given in reference [12], is strongly frequency dependent. Its relation to frequency is given by the expression

$$ j_a' = \frac{\omega^2 \theta^2 E_q}{1 + \omega^2 \theta^2}, $$

where θ is the time constant, q is the charge, and E is the electric field.

Thus, the electrothermal method allows determination of the value of ρ caused by the conductive current component, but it does not exclude detection of currents associated with surface effects if they occur.

The results of a study of the relationship $\rho / \rho_{200kcps} = F(f)$ for sand, clay, electrolyte with glass rods, and pure electrolyte are given in Fig. 90. The greatest decrease in resistivity with frequency, 85%, was found for the sand, and a lesser decrease, 25%, was found for the clay. The resistivity for the pure electrolyte and the electrolyte with glass rods did not change over the frequency range used. Ivanov explained the appearance of dispersion in resistivity by the presence of a double layer at the boundary between solid and liquid phases and comments on the possible relationship to the appearance of induced polarization.

In another study, surface polarization effects were avoided in the study of the frequency dependence of the dielectric constant by using mica strips as spacers between the electrodes and the sample [269]. Considering the use of such a noncontact method, Korennov and Chernyi obtained a large decrease (3-14 fold) at

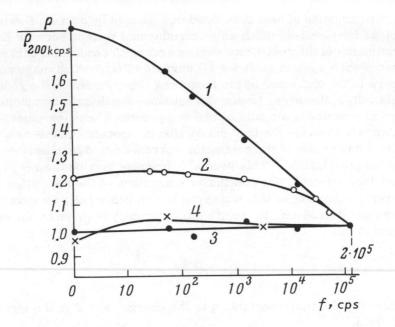

Fig. 90. Curves for the variation of relative resistivity [268]: (1) sand,
(2) shale, (3) electrolyte, (4) electrolyte with glass tubes.

50 kcps to 20 Mcps frequencies in the dielectric constant for sandstone, marl, and tuff (w = 1.16 to 3.7%). The variation, as was mentioned by the authors, depended mostly on the amount of water in the rock, and not very much on the petrology. An increase in the amount of dispersion in the dielectric constant was observed with increasing water content.

Combined studies of the dielectric constant and resistivity both for samples containing their original water contents and for dried samples of sandstone and siltstone at frequencies ranging from 50 cps to 30 Mcps have been carried out by Keller and Licastro [270]. In agreement with other authors, they found a large decrease in the resistivity of dry samples with increasing frequency, with ε remaining constant, and, in contrast, the absence of dispersion in resistivity in samples with a high water content, accompanied by a strong dispersion in ε (Fig. 91). Calculations of the loss tangent show that in wet rocks with a high dielectric constant, conduction is ohmic (tan δ > 1), while in samples with low values for tan δ, conduction is by displacement currents.

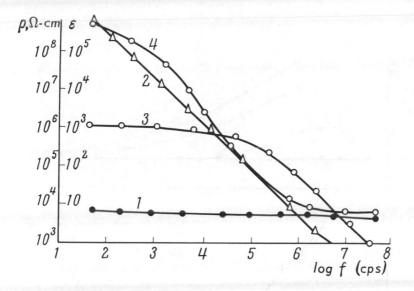

Fig. 91. Dispersion of the electrical properties of sandstone with frequency and moisture content (from Keller and Licastro): (1) dielectric constant of a dry sample, (2) resistivity of a dry sample, (3) resistivity of a wet sample, (4) dielectric constant for a wet sample.

In considering the possibility that the high values of ε found in wet rocks may be explained in terms of the Maxwell–Wagner theory for interfacial polarization, they conclude that such an explanation can be valid only for very large water contents. The authors explain the appearance of dispersion in the dielectric constant of rocks as the result of the presence of semiconducting clay particles, which give rise to a membrane effect, which causes a high value for ε at low frequencies as well as dispersion. In another paper, Howell and Licastro [14], in describing studies with pure sands and electrodialyzed clay fractions, arrive at similar conclusions and discuss the probable relationship between dispersion in dielectric constant and induced polarization.

The best discussion of the physical basis for the dispersion of the electrical properties for sedimentary rocks is that given in papers by Chelidze [27, 224]. Here, in contrast to foreign investigators [14, 270], he has considered the possible effect of electrode surface effects and devised methods for their exclusion, using a noncontact method with samples of different thicknesses. The results of this study apply mainly to sandstone, limestone,

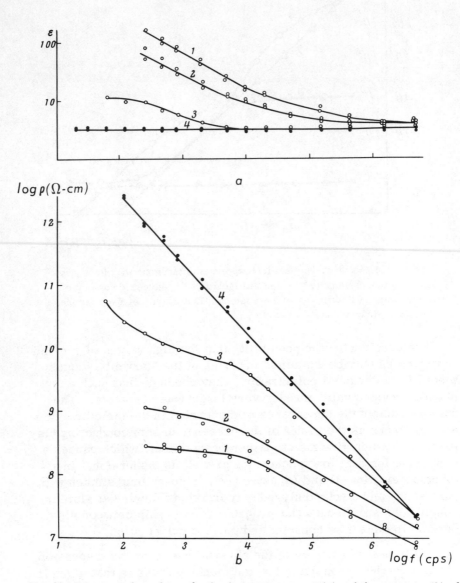

Fig. 92. Frequency dependence for the dielectric constant (a) and the resistivity (b) of a quartz sand. Water content in percent: (1) 12, (2) 0.4, (3) 0.03; (4) air dried with P_2O_5.

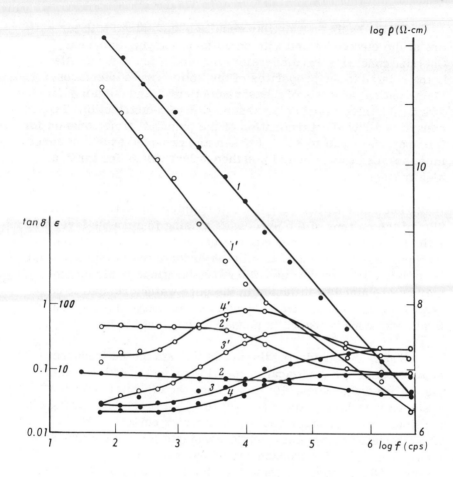

Fig. 93. Dispersion of the electrical properties of bentonite clay (size $200\,\mu$) at 28°C: (1) resistivity, (2) dielectric constant, (3) loss tangent, (4) the same in P_2O_5 dried atmosphere. The curves with open circles were measured with 10% moisture in the samples.

shale, quartz sand, and bentonite clay with different water contents over the frequency range 20 cps–50 Mcps. In the listed rock types, some degree of increase in dispersion of ε with increasing moisture content was established, as well as a decrease in the variation of ρ with frequency. These conclusions are substantiated by the data shown in Fig. 92. Chelidze indicated that a maximum is observed at frequencies of 10^4–10^5 cps for the loss tangent in most rocks, and at frequencies below 10^3 cps in a few rocks.

On the basis that similar results are obtained with air buffers at the electrodes and with samples of varying thickness, Chelidze concludes that dispersion in ε and ρ are volume effects, being related to the properties of the solid–liquid interfaces. Mathematical analysis of measurements of ε and the tan δ for "Chuneshi" clay based on an assumption of a distribution of time constants allowed determination of the principal time constant for this material equal to $8.5 \cdot 10^{-5}$ sec and provided excellent agreement between experimental and theoretical values for tan δ at high frequencies.

The same author [27], by determining the magnitude and sign of the entropy of the processes arising in a double layer when a field is applied, concluded that high-frequency relaxation polarization is caused by distortion in the orderliness of dipoles at the boundaries between phases. "Slow" relaxation, in his opinion, may be explained by diffusion in the pore water, related to concentration gradients. This hypothesis can not yet be considered final, and more research is needed to substantiate it.

Thus, in considering the available data on the frequency dependence of the electrical properties of rocks, it is apparent that at low frequencies, 10^2 to 10^4 cps, the dielectric constant assumes very large values ($\varepsilon = 10^3$ to 10^5) and exhibits strong dispersion for sedimentary rocks with a water content of 10–12%. The difference between dielectric constants for wet and dry samples is negligible for frequencies of the order of megacycles per second. The frequency dependence for resistivity as a function of water content is quite different. The less the water content and, thus, the higher the resistivity, the greater is the decrease in resistivity with increasing frequency. The rapid decrease in ε with a minor change in ρ contradicts theory, and, in our opinion, the existence of such a relation requires extensive verification. For this reason, Volarovich, Valeev, and I [223] have carried out a study of resistivity in sedimentary rocks with different water contents over the frequency range 30 to 20,000 cps. Resistivity measurements were made for limestone, marl, dolomite, sand marl, two types of siltstone, and three types of sandstone.

The results of these measurements indicated that little or no dispersion occurs in water-bearing rocks with a resistivity of 10^4 Ω-cm or less, but dispersion becomes more significant as the

water content is decreased, that is, as the resistivity is increased.
For example, for samples of sandy marl (w ≤ 1.87%) and siltstone
(w ≤ 2.48%) with resistivities of the order of 10^4 Ω-cm or less, no
dispersion in ρ occurs, but for the same rocks with moisture contents of 0.67 and 0.97%, respectively, the resistivity of the sandy
marl dropped from $2 \cdot 10^6$ to $9 \cdot 10^5$ with increasing frequency and
the resistivity of the siltstone dropped by more than an order of
magnitude – from $1.2 \cdot 10^6$ to $8 \cdot 10^4$ Ω-cm. The different rates
of change in ρ for these rocks probably are caused by differences
in petrography.

The constant value of ρ for rocks with relatively low resistivity may be explained as follows. The conductivity in rocks
with a high water content is contributed by the conductive liquid
phase. Therefore, the properties of the electrolyte should also
be the properties of the water-bearing rock. Theory and experimental data both indicate the absence of dispersion in resistivity
in electrolytes [268]. Thus, it is reasonable that there should be
no dispersion in the resistivity of rocks with large water contents.
Moreover, the absence of such dispersion in water-bearing sedimentary rocks may be explained using dielectric theory. Skanavi
[12] has shown that at low frequencies and with a large amount of
free conduction in a dielectric, that is, when $4\pi \delta_{free}/\omega \gg 1$, the
loss tangent is inversely proportional to frequency, while the resistivity does not depend on frequency. With the presence of a
relaxation process, the relaxation conduction current component
will be very small in comparison with the free conduction current,
and, therefore, a variation of j_a' with frequency will have practically no effect on the total current.

In our studies, with $f = 2 \cdot 10^4$ cps, the magnitude of $4\pi \delta_{free}/\omega$
is $0.9 \cdot 10^3$, that is, much greater than unity. Therefore, the
constant value of ρ for increasing frequency is entirely reasonable.

In explaining the origin and increasing importance of dispersion of resistivity in rocks with lower water contents, it is necessary to determine the roles of surface effects and polarization, as
well as collection of charge near electrodes and induced polarization. Also, it must be remembered that various relaxation processes become more likely with increasing rock resistivity.

We have made measurements on samples with various shapes
in order to evaluate the role of surface effects. These measurements have shown that a five- or sixfold change in length is not

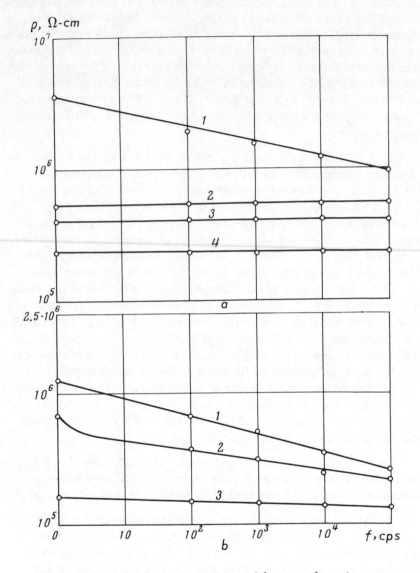

Fig. 94. Relationship between resistivity and frequency for various water contents in percent: a — dolomite (1) 0.36, (2) 0.63, (3) 0.85, (4) 1.7; b — sandstone (1) 0.58, (2) 0.61, (3) 0.68.

enough to allow positive identification of surface effects, but a difference in the nature of the frequency dependence of resistivity between large and small samples is found after an increase of tenfold or more in the thickness of a sample. For example, for a large sample of dolomite (h = 80 mm) with a water content of 0.63% and higher, the resistivity remained constant to a frequency of 10^5 cps (Fig. 94a), while the resistivity for a small sample (h = 3 mm) with the same water content dropped by a factor of 8 over the same frequency range. In the case of sandstones, as may be seen from Fig. 94b, the dispersion in resistivity may be seen somewhat sooner than in the case of dolomite. Therefore, petrography of the samples should be studied. At higher resistivities, greater than 10^6 Ω-cm, dispersion which did not depend on sample size was observed for all the sedimentary rock samples. Apparently, with increasing rock resistivity, the importance of relaxation polarization increases, possibly explained by the presence of inhomogeneities in the rock, contact effects at boundaries between the solid and liquid phases, and so on.

Thus, the available data indicate that in water-bearing rocks which have a resistivity of less than 10^5 Ω-cm in the natural state, dispersion in resistivity at frequencies of 10^2 to 10^5 cps is absent or very slight. At higher frequencies, according to the data of Keller and Licastro [270] and Chelidze [224], dispersion occurs and becomes more pronounced, but for resistivity in soils, variation in ρ does not take place below a frequency of 10^7 cps. The behavior of electric properties of low-resistivity rocks at high frequencies has been studied, but further work is needed.

No conclusion may be drawn concerning the possible frequency dependence of the dielectric constant. The very high values of the dielectric constant measured by a number of investigators for wet sedimentary rocks are subject to doubt in view of the possibility of surface polarization contributing to the results. In this respect, Valeev and I have carried out a study with various rocks with indirect contact between the electrodes and the samples using mica buffers. This prevented the electrochemical reactions which take place at the boundary between a metal and an electrolyte when current flows. Samples of sedimentary rocks on which measurements were made by the two methods exhibited a strong dispersion in dielectric constant for large contents (12-26%) over the frequency range 10^2-10^3 cps, the change being of the order 10^4 to 10^5 when

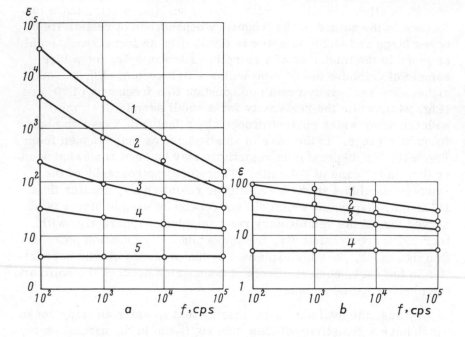

Fig. 95. Relationship between dielectric constant and frequency: a — siltstone with incomplete contact between the electrodes and the sample, moisture contents in percent (1) 15.7, (2) 12.0, (3) 5.1, (4) 2.1, (5) dry; b — sandstone with a mica separator between the electrodes and the sample, moisture contents in percent (1) 0.71, (2) 0.15, (3) 0.05, (4) dry.

electrodes were attached directly to the samples (Fig. 95a). At the same time, the dielectric constant measured for a siltstone sample with mica buffers was found not to depend on frequency, having a value of 35. Dispersion was significant when the water content was decreased to 2.1%, and was particularly strong for the same rock with a water content of 0.95%. No dispersion was found for the dried sample. Constant values of ε were also found for other dry rocks (Fig 95b) and for rocks saturated with oil. Using the same measurement technique, dispersion in ε and a relatively high value for the dielectric constant were obtained for an organic limestone, but much less than those obtained with direct contact between the electrodes and the samples. In the sandstone, the large value of ε is apparently caused by the presence of ore minerals.

Thus, our data show that it is necessary to exclude electrode polarization effects in measuring ε for water-bearing rocks. Measurements should be made on samples at high frequencies to provide corrections for the effect of direct contact between electrodes and a sample on the measured value of ε as well as for the effect of a mica buffer.

Investigations of the dielectric constant of water with different electrode separations, conducted by Efremov [271] and later by K.A. Valeev, indicated that electrode polarization apparently increased the observed value and enhanced the dispersion. This further substantiates the idea that direct contact between electrodes and a sample containing water leads to an erroneously large value of the dielectric constant and strong dispersion. However, this does not exclude the possible existence of volume effects at boundaries between the solid framework of a rock and the pore water and at electrical double layers. Conclusions concerning such questions drawn by Chelidze are quite reasonable but require further supporting experimental data, not only on the frequency dependence of dielectric constants in rocks caused by double layers, but also on induced polarization and the seismo-electric effect. It must be explained why the induced polarization and seismo-electric effects first increase with increasing water content and then decrease to low value at high water contents. At the same time, the dielectric constant for rocks measured with electrodes attached directly to the samples increases with undiminished intensity with increasing water content. When mica buffers are used or when measurements at high frequencies are made ($f \approx 10^6$ cps), a weaker increase in ε with water content is found.

At high frequencies, greater than 10^5 cps, the dielectric constant of rocks with moderate water contents ($w \leq 2\%$) does not decrease much with increasing frequency, while the variation is somewhat greater for higher water contents.

It has been established that the course of research in the area of frequency dependence of the dielectric constant must be directed toward development of a universally acceptable measurement technique or toward determining the nature of dispersion in ε by some indirect method for water-bearing rocks.

In the following section, we will consider the dielectric constant and its behavior in metamorphic and igneous rocks. Considering what we now know about the effect of electrode polarization, we should be particularly critical of the reported data for $\varepsilon = F$ (f) for these rocks.

Metamorphic and Igneous Rocks. According to the results of a number of studies [14, 27, 267, 272, 273], the electrical properties of metamorphic and igneous rocks exhibit dispersion similar to that in sedimentary rocks when water is present. However, the dielectric constant and its change with frequency are smaller in such rocks owing to their normally lower water content. Veshev [267] in studying room-damp samples of marble and granite over the frequency range from $5 \cdot 10^4$ to $5 \cdot 10^7$ cps has found a significant increase in conduction (about two orders of magnitude) and an insignificant change in ε for granite, and a weak increase in ε for marble. Data on the dispersion of the dielectric constant and resistivity given in another reference [272] for a broad range of frequencies indicates a good relationship ε, $\rho = F$ (f). The main shortcoming of the paper is the lack of quantitative data on water contents. Veshev considers that the existence of dispersion in the electrical properties of rocks is merely apparent, being caused by the increase in adsorption of current at higher frequencies of the incident electric field, which is not taken into consideration.

The most extensive data on the frequency dependence of the dielectric constant in dry igneous and metamorphic rocks at frequencies from 50 cps to 30 Mcps are given in references [14] and [254]. Igneous rocks of the types diabase, diorite, gabbro, and syenite, as may be seen in Fig. 96, are characterized by much more pronounced variations in ε than are granite and obsidian. Among the metamorphic rocks, talc slate exhibited the largest variation in ε with frequency. Dry rocks such as serpentinite, olivene pyroxenite, peridotite, weakly serpentinized peridotite, amphibolite, dunite, serpentine, pyroxenite, and diorite from the Kola Peninsula exhibit a weak frequency dependence for the dielectric constant at frequencies from 100 to 10^6 cps according to data obtained by Bondarenko [31]. A number of ultrabasic rocks maintained a high value of $\varepsilon = 12$ to 16 at frequencies up to 10^7 cps. It is supposed that this is explained not only by the basic mineral components, but also by the presence of ore minerals, because homogeneous rocks exhibit lesser values for ε.

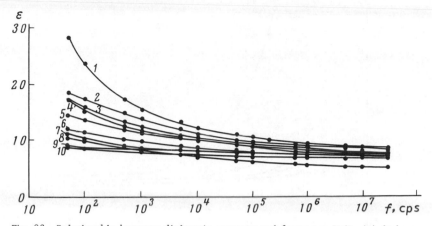

Fig. 96. Relationship between dielectric constant and frequency [14]: (1) diabase, (2) diorite, (3, 4) gabbro, (5) diabase, (6) syenite, (7) olivene, (8, 10) granite, (9) obsidian.

Most of the alkaline rocks from the Kola Peninsula are characterized by high values of the dielectric constant at 100 cps and a pronounced decrease in the dielectric constant at frequencies of 10^3–10^4 cps when water is extracted (Fig. 97). At high frequencies, the behavior is different, with the range of values for ε being relatively narrow, 5–10, and the curve for $\varepsilon = F(f)$ lies nearly parallel to the abscissa. Data on the ore mineral content and chemical composition are important in explaining the high values of ε.

With increasing water content, the dielectric constant for igneous rocks increases markedly, and a strong dependence on frequency becomes apparent. For a moisture content of 0.09%, ε decreases uniformly from 100 to 8 over the frequency range 10^3 to 10^7 cps. It may be supposed that such a pronounced effect by small amounts of water is caused by electrode polarization effects, which become relatively more important as the water content is increased.

The loss tangents for the investigated rocks from the Kola Peninsula decrease with increasing frequency, with the largest decrease being observed in alkaline rocks. The decrease in tan δ as a function of frequency persists over the entire frequency range from 10^2 to 10^7 cps, as may be seen in Fig. 98. A number of basic and ultrabasic rocks, as, for example, olivene pyroxenite,

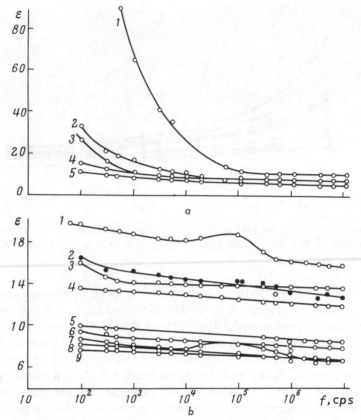

Fig. 97. Relationship between dielectric constant and frequency for
acidic, intermediate, and basic rocks: a — (1) leucophyre, (2) vein-
type melaphyre, (3) monocrystalline nepheline, (4) nepheline (under
vacuum), (5) the same after heating at 900°C; b — (1) plagioclase
peridotite, (2) serpentinite, (3) olivene pyroxenite, (4) peridotite, (5)
olivene pyroxenite, (6) amphibolite, (7) serpentinized dunite, (8)
weakly serpentinized dunite, (9) pyroxenite.

serpentinized dunite, and amphibolite, exhibit nearly constant values
of tan δ at frequencies beyond 10^4 cps, up to $f = 10^7$ cps and possibly
higher (Fig. 99).

Very good studies of polycrystalline pyrolusite from ore rock
have been carried out by Das [274] and others [25]. The primary
result was an anomalously high value for the dielectric constant, in
the range $2 \cdot 10^5$ to $8 \cdot 10^5$. The cause lies in the structure, which

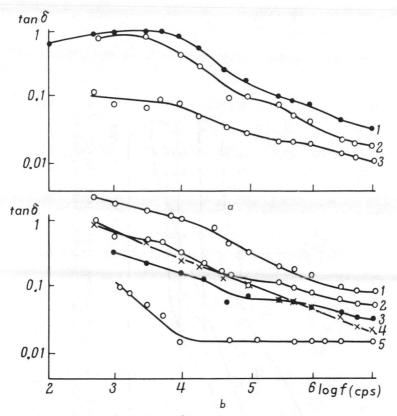

Fig. 98. Frequency dependence of tan δ in acidic rocks: a – (1) leucophyre, (2) vein-type melaphyre, (3) urthite; b – (1) uvite, (2) trachyte fayalite, (3) rischorrite (in vacuum), (4) aegerine, (5) nepheline trachytc (after heating at 900 °C).

is similar to the structure of rutile, which also has a large dielectric constant. Frequency dependence in the dielectric constant of pyrolusite appears as a decrease with increasing frequency, and the variation in tan δ is characterized by the existence of a maximum. With the acceptance of AC field methods in mining geophysics, it is important to develop studies of such ore minerals.

The dielectric properties of coals as a function of frequency have been studied only to a relatively small extent. The Japanese researchers Miyasita and Higasi [273] have studied ten grades of coal at room temperature and over the frequency range 300 cps–50

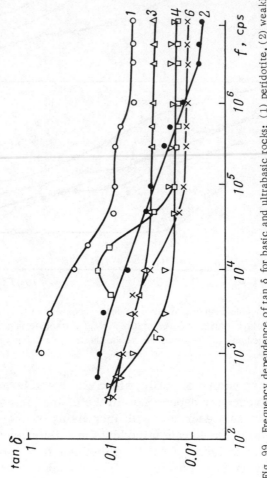

Fig. 99. Frequency dependence of tan δ for basic and ultrabasic rocks: (1) peridotite, (2) weakly serpentinized peridotite, (3) olivene pyroxenite, (4) serpentinized dunite, (5) olivene pyroxenite, (6) amphibolite.

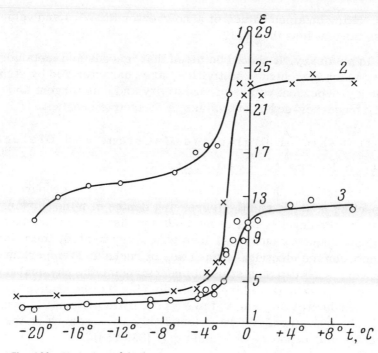

Fig. 100. Variation of dielectric constant with temperature in soils with various moisture contents: (1) compact clay with w = 35.5%, (2) fine-grained sand with w = 9% in surface-dried soil, (3) w = 3%; frequency 10^6 cps.

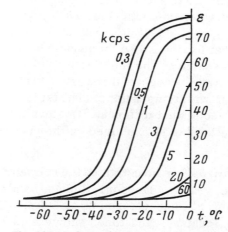

Fig. 101. Relationship between dielectric constant and temperature in ice at various frequencies.

Mcps. They obtained values of ε no greater than 5.5 and values
of loss tangent less than 0.1.

In summary, it should be noted that igneous and metamorphic
rocks, which have high resistivities, are characterized by strongly
frequency-dependent values of resistivity and loss tangent and
weakly frequency-dependent values of dielectric constant.

Relationship of Dielectric Constant and Dielectric
Loss in Rocks to Temperature

In relation with the development of subsurface geophysical
prospecting methods and the increasing depths of mine workings,
as well as the development of methods for disaggregating rocks
with high-frequency currents, it is necessary to study the relationship between the electrical properties of rocks and temperature.
Moreover, such data assist us in understanding the physical properties of the earth as a planet. Considering the increasing interest
in relationships of the form ε, tan δ = f (t), little material is available in the literature. We have found only a few papers on the
temperature dependence of ε and tan δ [33, 254].

Data on the behavior of the dielectric constant of rocks at negative temperatures is almost completely lacking. Dostovalov has
given only some preliminary results from measurements made at
temperatures from -20 to 8°C at a frequency of 10^6 cps for clay and
sand with various water contents (Fig. 100). According to the
curves for ε = f (t), the dielectric constant for sands at low temperatures is about that for ice. The temperature-dependent behavior
of ε in sand is the same as the function ε = f (t) for ice as shown in
Fig. 101, so that we may recognize three areas in the variation of the
dielectric constant with temperature. The temperature dependence,
ε = f (t), in clay is somewhat different. Dostovalov supposes that the
nature of the relationship is explained by the presence of unfrozen
water.

Bondarenko [254] has investigated the relationships ε, tan δ =
f(t) for quartzite, granite, sandsone, diabase, and syenite at a
frequency of 5 kcps at temperatures up to 800°C. For rocks including quartzite, quartz, sandstone, and vein quartz, he found the
same sort of variation in ε with temperature as is apparent in the
relations ε_1, ε_3 = f (t) for single crystals of quartz containing

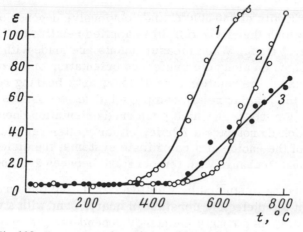

Fig. 102. Temperature dependence of the dielectric constant:
(1) sandstone, (2) quartzite, (3) vein quartz.

impurities. As may be seen in Fig. 102, the dielectric constant
remains constant over comparatively wide temperature ranges.
The temperature at which a rapid rise in the dielectric constants starts

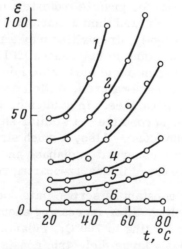

Fig. 103. Relationship between dielectric constant and temperature
for quartz sand at various frequencies (water content 0.4%): (1) 240
cps, (2) 480 cps, (3) 800 cps, (4)
2 kcps, (5) 5 kcps, (6) 5 Mcps.

falls in the limits of 250-600°C and, obviously, depends on the direction in which the electric field is applied relative to the optical and electrical axes. When measurements are made with the electric field directed along the preferred orientation of the optical or electrical axes, the relation $\varepsilon = f(t)$ for quartz-bearing rocks must agree with the similar relationship $\varepsilon = f(t)$ for the x-axis or z-axis in quartz. For rocks in which preferred orientation does not exist, that is, which do not have a fabric, or for molten samples, independently of the choice of a coordinate systems, the dielectric constant is found to change with temperature between 200 and 600°C.

According to studies by Chelidze [224], the temperature dependence of the dielectric constant of quartz sand with a water content of 0.4% is strongly frequency dependent. As may be seen in Fig. 103, with increasing frequency the curve $\varepsilon = f(t)$ flattens and at a frequency of 1.5 Mcps, over the indicated temperature range, the experimental points lie nearly parallel to the abscissa. Thus, the same sort of temperature and frequency dependence is found for rocks as is characteristic of many minerals with relaxation polarization.

Experimental data for granite indicate a minor change in ε for temperatures to 450°C and then a more rapid change with temperature [254]. Apparently, impurities play a major role in this behavior because, according to references [231, 243], the dielectric constants for phlogopite and muscovite micas remain constant over a broad range of temperatures, while ε for feldspars increases only slightly with temperatures. In studies of quartz-free rocks – syenite, diabase, and serpentine – a similar sharp bend in the curve $\varepsilon = f(t)$ has been found. Also, a much stronger increase in the value of ε has been observed in diabase and particularly in serpentine than is observed in quartz-bearing rocks.

The dielectric constant of serpentinite does not return to its original value on cooling, after increasing anomalously by five orders of magnitude on heating to 700°C. Polarization effects which can explain anomalously large dielectric constants in rocks have not been well-studied. Considering the available literature [275], we may assume that the anomalous change in ε for serpentinite may be caused by the presence of ore minerals, which are semiconductors. Of the rocks which have been investigated, pyrolusite also has a very large dielectric constant. According to references [25, 274], the temperature dependence of ε for pyrolusite is

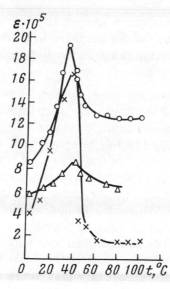

Fig. 104. Relationship between dielectric constant and temperature
for some samples of pyrolusite.

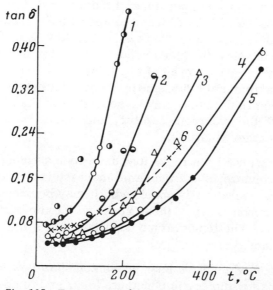

Fig. 105. Temperature-frequency relationships for
$\tan \delta$ in granite: (1) 50 kcps, (2) 150 kcps, (3) 500
kcps, (4) 1.5 Mcps, (5) 5 Mcps, (6) 8 Mcps.

characterized by a maximum around 40°C (Fig. 104). This temperature was taken by the investigator to be the Curie point. The
temperature behavior for tan δ in pyrolusite follows an exponential form.

Bondarenko [254] has also studied the temperature and frequency dependence of the loss tangent in granite and sandstone
(Fig. 105). A series of curves for tan δ =f (t) were obtained for
granite, shifting toward higher temperatures with increasing frequency, with the exception of one curve (6) obtained at a frequency
of 8 Mcps which shifted toward lower temperatures. The temperature-frequency relationship for tan δ for sandstone differed by
showing only minor changes for temperatures up to 400°C and then
a sharp increase in loss from a hundredth to some tenths on further
temperature rise. The increase in loss with increasing temperature may be explained superficially by conduction losses. However,
polarization processes apparently play some role, as well as some
type of ionic relaxation associated with weakly bonded ions, because the proper shift in the tan δ =f (t) curves is observed in the
case of granite.

Sevast'yanov et al. [276] has found a maximum value for ε of approximately 2.2 at a temperature of the order of 100°C and a decrease to a value of 1.5 at higher temperatures in measurements
made on bitumenous coal. The values 1.5 to 2.0 are somewhat
low in comparison with values reported by other investigators.
The temperature dependence of tan δ for coals is also characterized by a maximum at 100°C and a minor minimum at 200-300°C.
The maximum in the tan δ curve, according to the author, is associated with the evolution of water, and the minimum with the
expulsion of other volatiles.

Thus, with all the rocks that have been studied, except for
coal and pyrolusite, about the same increase in ε and tan δ with
increasing temperature is observed. It may be presumed that the
increase in conductivity of these rocks with temperature is the
major factor. The temperature dependence of ε and tan δ in rocks
should be studied further.

The data which have been presented show how poorly various
relationships for dielectric constant, loss, and temperature in all
sorts of rocks have been studied. Such data are extremely necessary in both geophysics and in mining.

Chapter V

Conclusions

The development of new and widely applied electrical exploration methods based on the use of electromagnetic fields for studying the structure of the earth's crust and upper mantle and the development of new methods for disaggregating rocks and economically valuable ores have brought to light a number of problems whose solution requires a detailed and complete study of the electrical properties of rocks. It is obvious that the measurement of electrical properties goes beyond the limits of geophysics. In one direction, it is impossible to arrive at reasonable conclusions about the electrical properties of rocks without knowledge of composition, texture, temperature, and pressure. The nature of changes in the electrical properties of rocks over a broad spectrum of frequencies is of great interest, as are correlations between electrical properties and other physical properties, such as elastic or magnetic properties.

The material contained in the present monograph has been taken from the literature and from the author's researches, and it suggests that the study of some of the relationships is still in a primitive stage, particularly the study of dielectric constant as a function of composition, water content, texture, temperature, and pressure. All of the work done in these areas is inadequate for arriving at any conclusions.

The most intensive studies reported in literature are those concerned with the resistivity of sedimentary rocks as a function of porosity. The results which have been obtained allow the quantitative interpretation of electric logs.

The studies of electrical resistivity in igneous rocks are inadequate. It would be particularly worthwhile to do further work on the resistivities of these rocks as a function of temperature and pressure.

This monograph contains the first discussion of dielectric loss in minerals and rocks. The subject could not have been covered earlier in view of the lack of data. Work in this area has only just been started. It would seem to be most important to study the frequency dependence of the loss tangent as well as the dielectric constant and resistivity in sedimentary and igneous rocks. The reality of high values of the dielectric constant, 10^4 or higher in water-bearing rocks at low frequencies, needs consideration. The answer to this question is quite important in deciding upon a universally acceptable method of measurement.

The development of new measurement techniques which would exclude the effects that occur at metal-rock contacts appears to be fundamental to further study of the electrical properties of rocks. There is also the possibility of studying electrical properties with indirect methods.

Finally, considering the broad application of AC fields in geophysical exploration, as well as in rock breakage by electrical methods, it is apparent that studies of the electrical properties of rocks must be taken more seriously.

Electrical properties in the broad sense also include piezo-electric, seismoelectric, and triboelectric effects, high voltage polarization, and induced polarization. Several of these effects are widely applied in geophysical exploration. They will be considered in a separate monograph.

The author is indebted to Prof. M.P. Volarovich, the Director of the Department of Physical Properties of Rocks, Institute of Physics of the Earth, Academy of Sciences of the USSR, for his views and aid in compiling this monograph, as well as to Dr. I.S. Zheludev, S.N. Kondrashov, Cand. G.A. Sobolev, and T.L. Chelidze for a number of critical comments.

References

1. F. Yu. Levinson-Lessing. Petrography. Moscow-Lenin-grad, Gosgeolizdat, 1940.
2. E. A. Kuznetsov. Petrography of Igneous and Metamorphic Rocks. Izd-vo MGU, 1956.
3. A. N. Zavaritskii. Igneous Rocks. Akad. Nauk SSSR, 1961.
4. R. O. Deli. Igneous Rocks and the Interior of the Earth. Moscow-Leningrad, ONTI-NKTP SSSR, 1936.
5. I. S. Shvetzov. Petrography of Sedimentary Rocks. Gosgeol-tekhizdat, 1958.
6. V. N. Kobranova and N. D. Leparskaya. Determining Physical Properties of Rocks. Gostoptekhizdat, 1957.
7. L. A. Shreiner. Physical Basis of Rock Mechanics. Mos-cow-Leningrad, Gostoptekhizdat, 1950.
8. N. B. Dortman, V. I. Vasilyeva, A. K. Veinberg, and others. Physical Properties of Rocks and Economic Minerals of the USSR. Nedra, 1964.
9. M. P. Volarovich. Investigation of physical-mechanical properties of rocks at high pressures. Geol. i.Geofiz., No. 4, 1961.
10. Z. I. Stakhovskaya. Investigation of the elastic and strength properties of rocks by the rupture method under high hydro-static pressure. Tr. Inst. Fiz. Zemli, No. 23 (190), 1962.
11. K. G. Orkin and P. K. Kuchinskii. Physics of Oil Beds. Gostoptekhizdat, 1955.
12. G. I. Skanavi. Physics of Dielectrics (Weak Fields). Mos-cow-Leningrad Gostekhteoretizdat, 1949.
13. H. G. Frohlich. Theory of Dielectrics. Oxford, 1949, [Rus-sian translation IL, 1960].
14. B. F. Howell and P. H. Licastro. Dielectric behavior of rocks and minerals. Am. Mineralog., Vol. 46, 1961.
15. J. C. Maxwell. A Treatise on Electricity and Magnetism, Vol. 1, Oxford, 1904.
16. K. Lichtenecker. Die Dielektrizitätskonstante natürlicher und künstlicher Mischkörpers. Phys. Z., Vol. 27, 1926.
17. V. I. Odelevskii, Calculation of the bulk conductivity of a heterogeneous system. Zh. Tekhn. Fiz., Vol. 21, 1951.
18. A. G. Tarkhov. On the relation of the dielectric constants of rocks to their mineral composition. Izv. Akad. Nauk SSSR, Ser. Geogr. i Geofiz., No. 2, 1947.

19.   A. S. Povarennykh. On the dielectric constant of minerals.
      From Sci. Works Krivorog. Min. Inst., Vol. 8, 1960.

20.   A. S. Povarennykh. On some basic questions in crystal
      chemistry and their application in mineralogy. Research
      Rept., All-Union Min. Inst., Ser. 2, Vol. 4, 1955.

21.   E. V. Rozhkova and L. V. Proskurovskii. Determination
      of the dielectric constants of minerals. From: Contempo-
      rary Methods of Mineralogic Studies of Rocks, Ores, and
      Minerals. Gosgeoltekhizdat, 1957.

22.   F. Birch et al. Handbook of Physical Properties. GSA
      Special Publ., 1940. [Russian translation IL, 1949].

23.   Rao Narayana. Dielectric constants of crystals. III. Proc.
      Indian Acad. Sci., A, Vol. 25, No. 2, 1947; Vol. 30, No. 2,
      1949.

24.   V. A. Koptsik and I. B. Kobyakov. Dielectric, piezoelec-
      tekhizdat, 1962.

25.   V. G. Bhide and R. V. Damle. Dielectric properties of
      manganese dioxide. Part I. Physica, Vol. 26, No. 1, 1960.

26.   V. A. Koptsik and I. B. Kobyakov. Dielectric, piezoelec-
      tric, and elastic properties of crystals of cancrinite. Kris-
      tollografiya, Vol. 4, No. 2, 1959.

27.   L. A. Chelidze and T. L. Chelidze. Concerning the question
      of the electrical properties of bentonite clay in ac fields.
      Tr. Kutai. Gos. Ped. Inst. im. A. Tsulukidze, Vol. 25, 1963.

28.   B. N. Dostovalov. Measurements of the dielectric constant,
      ε, and the resistivity, ρ, of rocks. Tr. Petrog. Inst. Akad.
      Nauk SSSR, No. 10, 1937.

29.   M. P. Volorovich, O. A. Tarasov, and A. T. Bondarenko.
      Investigation of the dielectric constant of rocks at atmospher-
      ic and uniaxial and hydrostatic (to 5000 kg/cm^2) pressures.
      Izv. Akad. Nauk SSSR, ser. geofiz., No. 7, 1961.

30.   N. K. Shchodro and N. M. Maslova. Determination of the
      dielectric constants of rocks and the effect of moisture in
      these rocks. Izv. Akad. Nauk SSSR, Otd. Matem. i Estest.
      Nauk, Nos. 6-7, 1935.

31.   A. T. Bondarenko. Frequency dependence of dielectric
      constant and loss tangent for igneous rocks from the Kola
      Peninsula. Izv. Akad. Nauk SSSR, Ser. Geofiz., No. 5,
      1965.

32.   B. N. Dostovalov. Measurement of the dielectric constant,
      ε, and resistivity, ρ, of rocks. Dokl. Akad. Nauk SSSR,
      Vol. 62, 3 (8), No. 2, 1935.

33.   B. N. Dostovalov. Electrical characterisitics of frozen
      rocks. Tr. Inst. Merzlotovedeniya (Permafrost) im. V. A.
      Obrucheva, Vol. 5, 1947.

34.   M. Groenewege, J. Schuyer, and D. W. Krevelen. Chem-
      ical structure and properties of coal. Dielectric constants
      of low rank and bituminous coals. Fuel, Vol. 34, No. 3, 1955.

35.   Y. Zyomoto. Dielectric characteristics of coals and pre-
      heated coals. J. Fuel. Soc. Japan, Vol. 37, No. 372, 1958.

36.   G. Ya. Chernyak. Methods of determining the Natural
      Moisture and Porosity of Sandy earth. Izd. Mos. Stroitel'stva
      Predpriyatii Metall. i Khim. Prom., Moscow, 1955.

37.   G. Ya. Chernyak. Dielectric Method for Determining the
      Moisture in Earth. Nedra, 1964.

38.   B. I. Korennov. Electrical properties of rocks and methods
      for their measurement in electromagnetic fields. From:
      Heat and Mass Volume in Frozen Soils and Rocks. Izd-vo
      Akad. Nauk, Moscow, SSSR.

39.   A. G. Tarkhov. On the resistivity (ρ) and dielectric con-
      stant (ε) of rocks in alternating electric fields. Geofizika,
      VSEGEI, Vol. 12, 1948.

40.   T. Rikitake. Electrical properties of soil at radio frequen-
      cies. Bull. Earthquake Res. Inst. Tokyo Univ., Pt. 2,
      Vol. 29, 1951.

41.   Rao Narayana. Dielectric anisotropy of rock. Proc. Indian
      Acad. Sci., Sect. A., Vol. 28, No. 1, 1948.

42.   F. D. Stacey. Dielectric anisotropy and fabric of rocks.
      Geofis. Pura Appl., Vol. 48, 1961.

43.   S. Mayburg. Pressure effect on the low frequency dielec-
      tric constant of ionic crystals. Phys. Rev., Vol. 79, 1950.

44.   A. C. Lyngh, and P. L. Parsons. Variation with pressure
      of the permittivity of polythene. Nature, Vol. 179, p. 686, 1957.

45.   J. Reitzel. Effect of pressure on the dielectric constant of
      vibreous silica. Nature, Vol. 178, No. 27, 1956.

46.   D. Gibbs and G. Hill. The variation of the dielectric con-
      stant of diamond with pressure. Phil. Mag., Vol. 9, No. 99,
      1964.

47. B. M. Vul and L. F. Vereshchagin. Relationship of the dielectric constant of barium titanate to pressure. Dokl. Akad. Nauk SSSR Vol. 48, No. 9, 1945.

48. I. P. Kozlobaev. Concerning the question of the dielectric constant of barium titanate. Dokl. Akad. Nauk SSSR, Vol. 104, No. 3, 1955.

49. V. Rudajev. Experimentelle Untersuchungen der Veränderungen der dielektrischen Konstante in Abhängigkeit von Druck. Sonderduck Freiberger Forschungsh. C., Vol. 126, 1960.

50. A. T. Bondarenko. Investigation of the dielectric constant of rocks at pressures to 50,000 kg/cm^2, and temperatures to 400°C. Izv. Akad. Nauk SSSR, Ser. Geogiz, No. 5, 1964.

51. M. P. Volarovich, A. T. Bondarenko, and É. I. Parkhomenko. Effect of pressure on the electrical properties of rocks. Tr. Inst. Fiz. Zemli Akad. Nauk SSSR, No. 23, 1962.

52. M. P. Volarovich, D. B. Balashov, and V. A. Pavlogradskii. Investigation of plasticity in igneous rock at pressure to 5000 kg/cm^2. Izv. Akad. Nauk SSSR, Ser. Geofiz, No. 5, 1959.

53. A. F. Ioffe. Physics of Semiconductors. Izd-vo Akad. Nauk SSSR, 1957 [translation, Academic Press Inc., New York, 1960].

54. P. E. Sarzhevskii. Relationship of electrical conductivity in quartz to electric field, temperature, and magnetic field. Dokl. Akad. Nauk SSSR, Vol. 32, No. 4, 1952.

55. B. M. Gokhberg. Electrical Conductivity of Dielectrics. GTTI, 1933.

56. A. R. Shul'man. Electrical conductivity of clays at high temperatures. Zh. Tekhn. Fiz., Vol. 10, No. 14, 1940.

57. A. P. Aleksandrov. Physics of Dielectrics, Vol. II. A. F. Vol'tera, ed., GTTI, 1932.

58. Ya. N. Pershits. Cathode effects in the electroconductive forming of dielectrics. From: Physics of Dielectrics. Akad. Nauk SSSR, 1958.

59. Ya. N. Pershits. On conditions for the formation and structure of inter-electrode layers in dielectrics. Zh. Eksperim. i Teor. Fiz., Vol. 28, No. 2, 1955.

60. B. M. Gokhberg and V. A. Ioffe. Investigation of electrical conductivity and high voltage polarization in crystal lattice. Zh. Eksperim. i Teor. Fiz., Vol. 1, 1931.

60a. B. M. Gokhberg, A. F. Ioffe, and K. D. Sinel'nikov. High-voltage polarization in dielectrics. Zh. Russ. Fiz.-Khim. Obshchestva, Vol. 105, No. 2, 1926.

61. A. M. Venderovich and B. Lapkin. Distribution of potentials in solid dielectrics. Zh. Eksperim. i Teor. Fiz., Vol. 9, No. 1, 1939.

62. A. M. Venderovich. High voltage polarization in lattices. Zh. Tekhn. Fiz., Vol. 23, No. 2, 1953.

63. M. S. Kosman and N. A. Petrova. Dielectric constant of rock salt at high temperatures. Izv. Akad. Nauk SSSR, Ser. Fiz., Vol. 22, No. 3, 1958.

64. V. A. Presnov. Concerning the question of the relationship of high voltage polarization in dielectrics to the voltage of the applied field. Zh. Tekhn. Fiz., Vol. 22, No. 6, 1952.

65. A. M. Venderovich. On the mechanism of high voltage polarization in NaCl type crystals. Zh. Tekhn. Fiz., Vol. 23, No. 2, 1953.

66. V. N. Lozovskii. Thermal ion polarization and slow processes in solid dielectrics. Izv. Akad. Nauk SSSR, Ser. Fiz., Vol. 22, No. 3, 1958.

67. S. A. Toporets. Methods of studying commerical coals. Izv. Akad. Nauk SSSR, Ser. Geofiz., No. 4, 1961.

68. N. S. Novosil'tsev and O. I. Prokopalo. Concerning the question of High Voltage Polarization and Methods for Its Measurement. Fiz.-Matem. Fak. Rostov Inst., Vol. 46, No. 7, 1959.

69. O. I. Prokopalo. On methods of measuring the potential of high voltage polarization. Fiz. Tverd. Tela, Vol. 2, No. 2, 1960.

70. B. V. Gorelik and V. T. Dmitriev. On the question of "true" electrical conductivity in solid dielectrics. Zh. Tekhn. Fiz., Vol. 18, No. 3, 1948.

71. N. V. Bogoroditskii and I. D. Fridberg. Concerning the question of electrical conductivity in solid dielectrics. Fizika Tverd. Tela, Vol. 6, No. 3, 1964.

72. A. M. Venderovich, F. I. Kolomitsev, and E. V. Sinyakov. Concerning the question of the nature of added conductance in dielectrics. Zh. Eksperim. i Teor. Fiz., Vol. 11, No. 4, 1941.

73.   Semiconductors in Science and Technology, Vol. 1. A. F.
       Ioffe, ed., Izd-vo Akad. Nauk SSSR, 1957.

74.   R. Keis. Effect of pressure on the electrical conductivity
       of indium antimonide. From: New Semi-Conducting Mate-
       rials. [Russian translation IL, 1958].

75.   L. F. Vereshchagin, A. A. Semerchan, S. V. Popova, and
       N. N. Kuzin. Variation of electrical resistivity of some
       semi-conductors at pressures to 300,000 kg/cm^2. Dokl.
       Akad. Nauk SSSR, Vol. 145, No. 4, 1962.

76.   Yu. N. Ryabinin, L. D. Livshits, and L. F. Vereshchagin.
       Concerning the question of the measurement of the electrical
       conductivity of silica under high pressure. Zh. Tekhn. Fiz.,
       Vol. 23, No. 7, 1958.

77.   L. F. Vereshchagin, A. A. Semerchan, and S. V. Popova.
       Investigation of the resistivity of cerium at pressures to
       250,000 kg/cm^2. Dokl. Akad. Nauk SSSR, Vol. 138, No. 5,
       1961.

78.   A. V. Fremke. Electrical Measurements. Moscow–Lenin-
       grad, Gosenergoizdat, 1954.

79.   N. G. Vostroknutov. Techniques for Measuring Electrical
       and Magnetic Magnitudes. Gosenergoizdat, 1958.

80.   K. B. Karandeev. Methods of Electrical Measurements.
       Moscow–Leningrad, Gosenergoizdat, 1952.

81.   F. Fritch. Electrical Measurements in Triaxial Conductors.
       Gostoptekhizdat, 1963.

82.   V. V. Panchenko, Temperature dependence of the dielectric
       constant in ionic crystals. Fiz. Tverd. Tela, Vol. 6, No. 2,
       1964.

83.   G. M. Zakharov, T. I. Nikitinskaya, and A. G. Khapachev.
       Impulse methods for measuring high resistances. Pribory i
       Tekhn. Eksperim., No. 4, 1960.

84.   N. A. Shitov, Meters for determining resistivity of rocks. Tr.
       Tsentral. Nauchn.-Issled. Gornorazved. Inst., No. 34, 1959.

85.   C. F. Rust. Electrical resistivity measurements on reser-
       voir rock samples by the two-electrode and four-electrode
       methods. Trans. Am. Inst. Mining and Metallurg. Engrs.,
       Vol. 195, 1952.

86.   V. N. Dakhnov. Promislovaya Geofizika. Gostoptekhizdat,
       1959. (also, Quart. Colo. School Mines, Vol. 57, No. 2,
       1962.)

87.   Handbook of Exploration Geophysics. Gosgeoncftcizdat, 1933.

88.   Handbook of Physical Constants, Moscow-Leningrad, ONTI-NKTP, 1937.

89.   J. Bardeen.   Electrical conductivity of metals.   J. Appl. Phys., Vol. 11, No. 2, 1940.

90.   M. V. Klassen-Neklyudova and T. A. Kontorova.   Properties of intercrystalline layers.   Usp. Fiz. Nauk, Vol. 22, No. 3-4, 1939.

91.   A. S. Semenov.   Resistivity of minerals with high conductivities.   Geofizika, Vol. 13, 1948 (VSEGEI), Gosgeolizdat.

92.   D. S. Parasnis.   The electrical resistivity of some sulphide and oxide minerals and their ores.   Geophys. Propecting, Vol. 4, No. 3, 1956.

93.   G. B. Abdulaev, G. M. Aliev, V. B. Anthonov, Ya. N. Nasirov, and A. A. Batshaliev.   Investigation of the thermal and electrical properties of ores with galena and chalcopyrite.   Izv. Akad. Nauk Azerb. SSR, Ser. Fiz.-Matem. i Tekhn. Nauk., No. 2, 1960.

94.   K. S. Krishnan and N. Ganguli.   Large anisotropy of the electrical conductivity of graphite.   Nature, Vol. 144, 1939, p. 667.

95.   A. K. Dutta.   Electrical conductivity of single crystals of graphite.   Phys. Rev. Vol. 90, No. 2, 1953.

96.   W. Primak and L. Fuchs.   Electrical conductivities of natural graphite crystals.   Phys. Rev., Vol. 95, No. 1, 1954.

97.   W. W. Tyler and A. C. Wilson.   Thermal conductivity, electrical resistivity and thermoelectric power of graphite.   Phys. Rev., Vol. 89, No. 4, 1953.

98.   G. H. Kunchin.   The electrical properties of graphite.   Proc. Roy. Soc., Vol. 217, No. 1128, 1953.

99.   A. K. Dutta.   Electrical conductivity of molybdenite crystals.   Nature, Vol. 159, 1947, p. 447.

100. R. Mansfield and S. A. Salam.   Electrical properties of molybdenite.   Proc. Phys. Soc. London, Ser. B, Vol. 66, 1953.

101. C. A. Dominicaly.   Magnetic and electric properties of natural and synthetic single crystals of magnetite.   Phys. Rev., Vol. 78, No. 4, 1950.

102. W. W. Piper. Some electrical and optical properties of
     synthetic single crystals of zinc sulphide. Phys. Rev.,
     Vol. 92, No. 1, 1953.
103. J. C. Marinace. Some electrical properties of natural
     crystals of iron pyrite. Phys. Rev., Vol. 96, No. 3, 1954.
104. Ya. Agaev and Kh. Érniyazov. Investigation of some elec-
     trical properties of pyrite. Izv. Akad. Nauk Turkm.SSR,
     Ser. Fiz., No. 5, 1963.
105 F. C. Smith. The pyrite geothermometer. Econ. Geol.,
     Vol. 42, No. 6, 1947.
106. G. A. Gorbatov. Thermoelectric properties of pyrite and
     galena and their possible relation to the temperature of
     mineral formation. From: Material on the Investigation of
     Mineral Oils, Gosgeoltekhizdat, 1957.
107. G. A. Gorbatov and L. N. Indolev. Research on the use of
     statistical analysis of thermoelectric force in galena for a
     preliminary zoning of ore formation. Yakutsk Filial,
     S. O., Akad. Nauk SSSR, No. 7, 1962.
108. E. V. Frantsesson. Thermoelectric properties of natural
     solutions, Geolog. i Geofiz., No. 3, 1963.
109. M. Telkes. Thermoelectric power and electrical resistivity
     of minerals. Am. Mineralogist, Vol. 35, No. 7-8, 1950.
110. K. Noritomi. Investigations of thermoelectricity for metal-
     lic and silicate minerals. Sci. Rept. Tohoku Univ., Ser.
     Geophys., Vol. 7, No. 2, 1958.
111. N. P. Bogoroditskii, V. V. Pasynkov, and B. M. Tareev,
     Electrotechnical Material, 3rd ed. Moscow-Leningrad
     Gosenergoizdat, 1955.
112. A. I. Zaborovskii. Electrical Exploration. Gostoptekhizdat,
     1943.
113. V. V. Skorokhod. On the electrical conductivity of disperse
     mixtures of conductors with non-conductors. J. Eng. Phys.,
     Vol. 2, No. 8, 1959.
114. I. K. Ovchennikov and G. G. Kilukova. Effective electrical
     conductivity of media with inclusions. Izv. Akad. Nauk SSSR,
     Ser. Geofiz., No. 1, 1955.
115. I. K. Ovchennikov. On the theory of effective electrical
     conductivity, magnetic permeability and dielectric constant
     of media with ore mineral inclusions. Tr. Vses. Inst.
     Razved. Geofiz., No. 3, 1950.

116. A. S. Semenov. Effect of texture on the resistivity of aggregates. Geofizika (VSEGEI), Vol. 12, 1948.

117. A. S. Semenov. Resistivity of ores and rock and the total evaluation of ore bodies as an object of electrical exploration. Tr. Vses. Nauchn. Issled. Inst. Razved. Geofiz., Vol. 1, 1949.

118. D. F. Murashov, E. V. Berengarten, A. V. Echeistova, and L. D. Khudyakova. Electrical Conductivity of Ore and Rock. Leningrad, Goelkom, 1929.

119. A. V. Bukhnikashvili. Electrical conductivity of ore and rocks of the Caucasus. Tr. Inst. Geofiz., Akad. Nauk Gruz SSR, Vol. 17, 1958.

120. Sh. M. Chkhekneli. Study of the electrical resistivity of rocks and ores. Tr. Inst. Fiz. i Geofiz., Akad. Nauk Gruz SSR, Vol. 11, 1949.

121. Kh. I. Amirkhanov. Investigation of the electrical conductivity of rocks. Tr. Azerbaid Filial, Akad. Nauk SSSR, Ser. Fiz.-Matem., Vol. 28, 1936.

122. A. V. Bukhnikashvili. Electrical resistivity of rocks and ore. Izv. Gruz. Indust. Inst., No. 12, 1940.

123. P. Mandel, J. W. Berg, and K. C. Cook. Resistivity studies of metalliferous synthetic cores. Geophysics, Vol. 22, No. 2, 1957.

124. R. R. McEuen, J. W. Berg, and K. C. Cook. Electrical properties of synthetic metalliferous ore. Geophysics. Vol. 24, No. 3, 1959.

125. E. I. Parkhomenko and A. T. Bondarenko. Effect of uniaxial pressure on the electrical resistivity of rocks. Izv. Akad. Nauk SSSR, Ser. Geofiz., No. 2, 1960.

126. V. A. Marinim. Behavior of rocks in Constant Electric Fields. Leningrad, 1938. Dissertation Leningr. Gos. Inst.

127. V. V. Kebuladze. Electrical conductivity of samples of rocks and ore from an antimonite deposite. Tr. Inst. Fiz. i Geofiz., Akad. Nauk Gruz SSR, Vol. 11, 1949.

128. A. A. Agroskin. Thermal and Electric Properties of Coals. Metallurgizdat, 1959.

129. A. P. Kovalev. Dry Method of Separating Coal. MÉI, 1946.

130. V. N. Dakhnov. Electrical Exploration in Oil and Gas Fields. Gosgeolizdat, 1951.

131. A. A. Agroskin and I. G. Petrenko. Electrical conductivity of slate and coal on heating. Izv. Akad. Nauk SSSR, Ser. Geol. No. 1, 1950.

132. N. N. Shumilovskii. Electrical characteristics of coal. Avtomat. i Telemekhan. No. 4, 1932.

133. A. A. Agroskin. Physical Properties of Coals. Metallurgizdat, 1961.

134. S. A. Toporets. On the effect of metamorphism on the electrical and elastic properties of commercial coals. Dokl. Akad. Nauk SSSR, Vol. 140, No. 2, 1961.

135. S. I. Vosanchuk. Electrical resistivity of Donbas coals. Sci. Works Lvov. Gos. Inst., Ser. Neft., Vol. 35, No. 6, 1955.

136. S. A. Toporets. On the electrical conductivity of petrographic ingredients of coals. Dokl. Akad. Nauk SSSR, Vol. 118, No. 1, 1958.

137. S. A. Toporets. Effect of composition of mineral impurities on the electrical conductivity of coal. Dokl. Akad. Nauk SSSR, Vol. 122, No. 2, 1958.

138. V. V. Grechukhin. Methods of determining the relative and the reduced apparent resistivity and their use for studying the quality of coal layers. Geofiz. Razvedka, No. 15, 1964, Nedra.

139. V. N. Dakhnov. Interpretation of the Results of Geophysical Investigations of Well Sections. Gostoptekhizdat, 1962.

140. L. P. Dolina. Determination of porosity, permeability and oil saturation from geophysical data, and research on their utilization in calculating oil production. From: Voprosi Neftepromislovoi Geologii, No. 12, Gostoptekhizdat, 1959.

141. G. S. Morozov. Method of determining porosity, permeability and specific surface of water saturated rocks from electric log data. Prikl. Geofiz., Vol. 19, 1958.

142. R. P. Éldzher. Investigation of carbonate rocks with well logging methods. Promys. Geofiz., No. 1, 1959.

143. V. N. Dakhnov, V. N. Kobranova, M. G. Latishova, and V. A. Ryapolova. Electrochemical well logging. Promys. Geofiz., 1956.

144. S. G. Komarov. Determining rock porosity from resistivity. Prik. Geofiz. Vol. 14, 1952.

145. F. M. Perkins, D. S. Osoba, and K. Kh. Raib. Relationship between resistivity of sandstones and the character of water

distribution in the pores. Promys. Geofiz., No. 1, 1959 [Translated from English].

146. I. E. Éidman. Electrical resistivity.   Prikl. Geofiz. Vol. 15, 1956.

147. G. E. Archie. Classification of porous carbonate rocks and their petrographic properties. Vopr. Promyslovoi Geofiz., 1957 [Translated from English].

148. R. F. Nil'sen. On the effect of tortuosity of pore channels on permeability and resistivity of rocks. Vopr. Promyslovoi Geofiz., 1957.

149. V. O. Winsauer, Kh. M. Sherin, P. Kh. Masson, and M. Vil'yams Relationship between pore geometry and resistivity of sand saturated with water. Vopr. Promyslovoi Geofiz., 1957 [Translated from English].

150. M. R. Wyllie, and A. R. Gregory. Formation factors in uncemented media: effect of grain form and cementation. Vopr. Promyslovoi Geofiz., 1957 [Translated from English].

151. A. I. Krinari. On the relationship between resistivity and reservoir properties of water-bearing clastic rocks. Geol. Nefti, No. 7, 1958.

152. Z. I. Keivsar. On the relation of formation factor to porosity, specific surface and permeability of rocks. Prikl. Geofiz., Vol. 19, 1958.

153. A. T. Boyarov. On the relation of the formation factor of a rock to the concentration of the saturating solution. Geologiya Geokhimiya, Geofizika., Vol. 1, 1960.

154. L. DeWitte. Determining saturation and porosity of shaly sands from well log data. Promyslovaya Geofiz., No. 1, 1959 [Translated from English].

155. V. N. Dakhnov and L. P. Dolina. Geophysical Methods for Studying Oil Reservoirs. Gostoptekhizdat, 1959.

156. B. Yu. Vendel'shtein. On the relation between formation factor, coefficient of surface conductance, diffusion adsorption properties of clastic rocks. Tr. MINKh i GP, Vol. 31, 1960.

157. Electrokinetic Properties of Capillary Systems. Moscow-Leningrad, Akad. Nauk SSSR, 1956.

158. G. V. Keller. Effect of wettability on the resistivity of sands. Vopr. Promyslovoi Geofiz., 1957 [Translated from English].

159. P. T. Kotov. Resistivity of rocks containing emulsions. Prikl. Geofiz., Vol. 39, 1964.

160. G. S. Komarov and Z. I. Keivsar. Determining permeability of oil bearing layers by resistivity, Prikl. Geofiz., Vol. 20, 1958.

161. V. N. Dakhnov and V. N. Kobranova. Study of the reservoir properties and oil saturation of productive horizons in oil fields with well logging data. Promys. Geofiz., 1952.

162. A. S. Gritsaenko. On electrical macro-anistropy in layered media. Research Reports, Saratov. Inst., Vol. 65, 1959.

163. A. A. Ananyan and V. N. Dobrovol'skii. On the electrical conductivity of frozen rock. Geol. i Geofiz., Vol. 3, 1961.

164. I. I. Goryunov. Electrical resistivity of fractured rock. Prikl. Geofiz, Vol. 38, 1964.

165. A. M. Nechai. Question of the quantitative evaluation of secondary porosity in fractured oil and gas reservoirs. Prikl. Geofiz., Vol. 38, 1964.

166. I. I. Goryunov and L. A. Molotkov. Electrical properties of fractured rocks and their relation to reservoir properties. Tr. VNIGRI, Vol. 165, 1961.

167. E. M. Smekhov. Laws of Development of Fractures in Rock and Fractured Reservoirs. Gostoptekhizdat, 1961.

168. A. M. Nechai. Evaluation of productivity and reservoir properties in fractured carbonate rocks. Prikl. Geofiz., Vol. 26, 1960.

169. L. S. Polak and M. B. Rapoport. On the relation between electrical and elastic properties in sedimentary rocks. Prikl. Geofiz., Vol. 15, 1956.

170. L. S. Polak and M. B. Rapoport. On the relation of the velocity of elastic waves to some physical properties of sedimentary rocks. Prikl. Geophys., Vol. 29, 1961.

171. O. B. Ukleba. Research in determining the silica content in rocks from electrical resistivity. Tr. Inst. Geofiz., Akad. Nauk Gruz. SSR, Vol. 21, 1963.

172. B. Gutenberg. Physics of the Earth's Interior [Russian translation IL, 1963]. Academic Press Inc., New York.

173. A. T. Akimov. Electrical resistivity of frozen ground. Dokl. Akad. Nauk SSR, Vol. 16, issue 8, 1937.

174. A. N. Zil'berman, A. P. Lubimov, and A. P. Bazhenova. Data on the freezing and thawing of soils by the electrical conductivity method. Geofizika, Vol. 5, issue 1, 1935.
175. P. I. Andrianov. Freezing Temperatures of Soil. Akad. Nauk SSSR, 1963.
176. M. A. Nesterova and L. Ya. Nesterov. Electrical conductivity of rocks below zero. Geofizika, Vol. 11, 1947 (Materiali VSEGEI), Gosgeolizdat.
177. M. S. Grutman. Experimental studies of the base of frozen layers in thick soil. Tr. Ukr. Nauchn.-Issled. Inst. Cooruzhenii, 1948.
178. A. A. Ananyan. Electrical conductivity of frozen rock and its relation to the freezing process and phase transitions in water. Tr. Sobeshch. Po Inzh. Geol. Svoistvam Gornykh Porod i Metodam Ikh Izucheniya, Vol. 2, 1957.
179. A. A. Ananyan. Relationship of the electrical conductivity of frozen rock to moisture content. Izv. Akad. Nauk SSSR, Ser. Geofiz., No. 12, 1958.
180. M. Mikhailov and G. Soya. On the temperature dependence of the electrical properties of some dielectrics. Zh. Tekhn. Fiz., Vol. 6, No. 5, 1936.
181. É. I. Parkhomenko and A. T. Bondarenko. Electrical conductivity of rocks at high temperatures under uniaxial pressure. Tr. Inst. Fiz. Zemli Akad. Nauk SSSR, No. 23, 1962.
182. H. P. Coster. The electrical conductivity of rocks at high temperatures. Monthly Notices Roy. Astron. Soc., Geophys. Suppl., Vol. 5, No. 6, 1948.
183. E. A. Polyakov. Study of the electrical resistivity and density of aqueous solutions of salt at high pressure and temperature. Prikl. Geofiz., Vol. 41, 1965.
184. K. Noritomi. Studies on the change of electrical conductivity with temperature of a few silicate minerals. Sci. Rept. Tohoku Univ., Ser. 5, Vol. 6, No. 2, 1955.
185. K. Noritomi and A. Asada. Studies on the electrical conductivity of a few samples of granite and andesite. Sci. Rept. Tohoku Univ., Ser. 5, Vol. 7, No. 3, 1956.
186. K. Noritomi. The electrical conductivity of rocks and the determination of the electrical conductivity of the earth's interior. J. Mining Coll. Akita Univ., Ser. A, Vol. 1, No. 1, 1961.

187. S. K. Runcorn and D. C. Tozer. The electrical conductivity of olivine at high temperatures and pressures. Ann. Geophys., Vol. 7, No. 11, 1955.

188. T. Murase. Viscosity and related properties of volcanic rocks at 800 to 1400°C. J. Fac. Sci., Hokkaido Univ., Ser. VI, Vol. 1, No. 6, 1962.

189. U. I. Moiseenko and V. E. Istomin. Investigation of the electrical conductivity of rocks at high temperatures. Geol. i Geofiz., No. 8, 1963.

190. E. B. Lebedev and N. I. Khitarov. Beginning of melting in granite and the electrical conductivity of its melts in relation to pressure of pore water. Geokhim., No. 3, 1964.

191. F. S. Zakirova. Study of the resistivity of minerals and rocks with their solutions. Dokl. Akad. Nauk SSSR, Vol. 154, No. 6, 1964.

192. T. Nagata. Some physical properties of the lava of volcanoes Asama and Mihara. Bull. Earthquake Res. Inst., Vol. 15, 1937.

193. A. N. Tikhonov, N. V. Lipskaya, N. A. Deniskin, N. N. Nikiforova, and Z. D. Lomakina. On electromagnetic sounding of deep layers in the earth. Dokl. Akad. Nauk SSSR, Vol. 140, No. 3, 1961.

194. S. Jain. Electrical resistivity of the crust and upper mantle at Eskdalemuir, South Scotland. Nature, Vol. 203, No. 4945, 1964.

195. L. M. Marmorshtein, I. M. Petukhov, B. B. Nersesyantz, and G. I. Morozov. Effect of the composition of cementing material and porosity on the variation of conductivity of sedimentary rocks under pressure. Tr. Soveshchaniya po Fizicheskim Metodam Issledovaniya Osadochnykh Porod i Mineralov, 1961.

196. L. M. Marmorshtein. Effect of pressure on the Resistivity and Formation Factor of Rocks. Dissertation, Nauchno-Issled. Inst. Geologii Arktiki, Leningrad, 1962.

197. A. L. Gonor and L. M. Marmorshtein. Method of computing the porosity of rocks under pressure. Tr. Inst. Geologii Arktiki, Vol. 132, No. 4, 1962.

198. I. M. Petukhov, I. M. Marmorshtein, and G. I. Morozov. On the use of variations in conductivity of rocks in studying their directional properties in place and reservoir properties. Tr. Vses. Nauchn.-Issled. Marksheiderskogo Inst., Vol. 42, 1961.

199. L. M. Marmorshtein. Effect of pressure on the physical properties of rocks. Tr. Inst. Geologii Arktiki, Vol. 132, No. 4, 1962.

200. V. M. Dobrynin. Variation of the physical properties of sandstones under hydrostatic pressure. Promyslovaya Geofizika, 1963.

201. I. F. Glumov and V. M. Dobrynin. Variation of the resistivity of water saturated rocks under the influence of overburden and formation pressure. Prikl. Geofiz. Vol. 33, 1962.

202. L. I. Orlov and R. S. Gimalaev. Effect of pressure on the resistivity of carbonate rocks. Prikl. Geofiz., Vol. 33, 1962.

203. J. Fatt. Effect of overburden and reservoir pressure on electric logging formation factor. Bull. Am. Assoc. Petrol. Geologists, Vol. 41, No. 11, 1957.

204. D. Wyble. Effect of applied pressure on the conductivity of sandstones. J. Petrol. Technol., No. 11, 1958.

205. C. R. Granville. Laboratory study indicates significant effect of pressure on resistivity of reservoir rock. J. Petrol. Technol. No. 4, 1959.

206. H. Hughes. The pressure effect on the electrical conductivity of peridotit. J. Geophys. Res., Vol. 60, No. 2, 1955.

207. M. P. Volarovich and A. T. Bondarenko. Investigation of the resistivity of rock samples under hydrostatic pressure to 1000 kg/cm^2. Izv. Akad. Nauk SSSR, Ser. Geofiz., No. 7, 1960.

208. É. I. Parkhomenko and A. T. Bondarenko. Investigation of the resistivity of rocks at pressures to 40,000 kg/cm^2 and at temperatures to 400°C. Izv. Akad. Nauk SSSR, Ser. Geofiz., No. 12, 1963.

209. M. P. Volarovich and D. B. Balashov. Investigation of the velocity of elastic waves in rock samples under pressures to 5000 kg/cm^2. Izv. Akad. Nauk SSSR, Ser. Geofiz., No. 3, 1957.

210. P. T. Kozyrev. Relationship of the conductivity of selenium to pressures of 30,000 atm. Fiz. Tverd. Tela, Vol. 1, No. 1, 1959.

211. U. I. Moiseenko, Istomin, and Ushakova. Effect of uniaxial pressure on the resistivity of rocks. Dokl. Akad. Nauk SSSR, Vol. 154, No. 2, 1964.

212. U. I. Moiseenko and V. E. Istomin. Resistivity of rocks
     at high temperature and pressure. Dokl. Akad Nauk SSSR,
     Vol. 154, No. 4, 1964.

213. R. S. Bradley, A. K. Jamil, and D. C. Munro. Electrical
     conductivity of fayalite and spinel. Nature, Vol. 193,
     No. 4819, 1962.

214. E. S. Itskevich, É. Ya. Atabaeva, and S. V. Popova. Effect
     of pressure on the conductivity of bismuth selenide. Fiz.
     Tverd. Tela, Vol. 6, No. 6, 1964.

215. A. I. Likhter and T. S. D'yakonov. Measurement of the rela-
     tionship of Hall effect in n-type germainium to pressures up
     to 1000 kg/cm^2. Fiz. Tverd. Tela, Vol. 1, No. 1, 1959.

216. E. S. Itskevich, S. V. Popova, and É. Ya. Atabaeva.
     Effect of pressure on the conductivity of bismuth telluride,
     Dokl. Akad. Nauk SSSR, Vol. 153, No. 2, 1953.

217. A. R. von Hippel, Dielectrics and Waves [Russian Trans-
     lation IL, 1960]. The Technical Press of MIT, 1954.

218. V. M. Brown. Dielectric [Russian translation, IL, 1961].

219. A. R. von Hippel. Dielectric Materials and Applications
     [Russian translation, 1959]. The Technical Press of MIT,
     1954.

220. O. V. Mazurin. Electrical Properties of Glass. Leningrad,
     Goskhimizdat, 1962.

221. K. A. Vodop'yanov and B. T. Karov. Effect of electrode
     material, thermal and electrolytic working on the dielectric
     properties of steatite ceramic. Izv. Vysshikh Uchebn.
     Zavedenii Fiz., No. 3, 1964.

222. A. A. Brant. Investigation of Dielectrics at Very High
     Frequencies. Moscow, Fizmatgiz, 1963.

222a. F. B. Gul'ko and A. I. Usherenko. On the question of
     determining the dielectric constant of solid dielectrics.
     Scientific-Technical Department, Moscow Institute of
     Energics, No. 9, 1956.

223. M. P. Volarovich, K. A. Valeev, and É. I. Parkhomenko.
     Resistivity of rocks in constant and alternating electric
     fields. Izv. Akad. Nauk SSSR Fiz. Zemli, No. 5, 1965.

224. T. L. Chelidze. Concerning the question of the frequency
     dependence of the electrical properties of rocks. Tr. Inst.
     Geofiz., Akad. Nauk Georgian SSR, Vol. 21, 1963.

225. L. Hartshorn. Radio-frequency measurements by bridge and resonance methods. London, Chapman, Hale, New York, 1947.

226. B. Hague. Alternating Current Bridge Methods. London, Pitman, 1946.

227. K. M. Sobolevskii and Yu. A. Shanola. Protected AC Bridges. Akad. Nauk SSSR, 1957.

228. K. B. Karandeev. Special Methods for Electrical Measurements. Moscow—Leningrad, Gosenergoizdat, 1963.

229. D. Karo. Electrical measurements and the calculation of the errors involved, Pt. II. London, Macdonald, 1953.

230. B. M. Tareev. Electrical Engineering Topics. Moscow-Leningrad, Gosenergoizdat, 1960.

231. K. A. Vodop'yanov. Concerning the question of dielectric loss in crystals at high frequency. Tr. Sibirsk. Fiz. Tekhn. Inst., No. 24, 1947.

232. K. A. Vodop'yanov. Dielectric properties of mica. Elektrichestvo, No. 11, 1950.

233. K. A. Vodop'yanov. Concerning the question of dielectric loss in mica at high frequency. Tr. Soveshch. po Tverdym Dielektrikam, 91, 1955.

234. K. A. Vodop'yanov. Temperature and frequency dependence of the dielectric loss angle in crystals with polar molecules. Dokl. Akad. Nauk SSSR, Vol. 84, No. 5, 1952.

235. K. A. Vodop'yanov. On relaxation dielectric loss in crystals with polar molecules at high frequency. Zh. Tekhn. Fiz., Vol. 24, No. 1, 1954.

236. K. A. Vodop'yanov and I. G. Vorozhtsova. Dielectric loss in muscovite mica with mineral inclusions of limonite and biotite at high frequency. Izv. Akad. Nauk SSSR, Ser. Fiz., No. 3, 1958.

237. A. P. Izergin. Temperature dependence of the dielectric loss angle and dielectric constant in phlogopite mica at industrial frequencies. Tr. Sibirsk. Fiz.-Tekhnol. Inst., No. 24, 1947.

238. A. P. Izergin. Dielectric loss angle and dielectric constant in phlogopite mica at industrial frequencies. Tr. Sibirsk. Fiz.-Tekhnol. Inst., No. 28, 1949.

239. J. Keymeulen. Dielectric measurements on clay minerals. Naturwissenschaften, Vol. 44, 1957.

240. A. V. Mal'tsev and V. F. Berdinskii. Temperature de-
     pendence of the dielectric constant of muscovite mica at
     frequencies from 50 cps to 1 mcps. Uchenye zapiski LGPI,
     207, 1961.
241. I. G. Vorozhtsova. Concerning the question of the nature of
     dielectric loss in mica. Fizika, No. 1, 1962.
242. K. A. Vodop'yanov. Dielectric loss in some crystalline
     materials at high frequency. Zh. Tekhn. Fiz., Vol. 19,
     No. 9, 1949.
243. K. A. Vodop'yanov. Investigation of the dielectric prop-
     erties of crystalline gypsum at high frequencies. Tr.
     Sibirsk. Fiz.-Tekhnol. Inst., No. 28, 1949.
244. N. A. Rodionova. Dielectric loss in $NiSO_4 \cdot 6H_2O$. From:
     Frizika Dielektrikov, Akad. Nauk SSSR, 1960.
245. K. A. Vodop'yanov, G. A. Galibina. Dielectric loss in
     pure and impure alkaline halide crystals at high frequency.
     Izv. Akad. Nauk SSSR, Ser. Fiz., Vol. 22, No. 3, 1958.
246. M. P. Tonkonogov. Dielectric relaxation in poly-crystalline
     solid dielectrics at high frequency. Izv. Tomsk. Politekhn.
     Inst., Vol. 91, 1956.
247. B. N. Matsonashvili. Dielectric constant, dielectric loss
     and conductivity in alkali-halide single crystals. Izv. Akad.
     Nauk SSSR, Ser. Fiz., Vol. 22, No. 3, 1958.
248. M. S. Chmutin. Relation of the dielectric constant of
     crystals of NaCl to temperature. Zh. Éksperim. i Teor.
     Fiz., Vol. 26, No. 5, 1954.
249. J. Keymeulen. Dielectric behaviour of kaolinite and dickite.
     J. Chem. Phys. Vol. 25, No. 2, 1956.
250. J. Keymeulen and W. Dekeyser. Dielectric loss of and
     defects in clay minerals. J. Chem. Phys., Vol. 27, No. 1,
     1957.
251. N. P. Bogoroditskii and I. D. Fridberg. Electro-Physical
     Basis for High-Frequency Ceramics. Gosenergoizdat, 1958.
252. M. Stuarts. Dielectric constant of quartz as a function of
     frequency and temperature. J. Appl. Phys., Vol. 26,
     No. 12, 1955.
253. V. G. Zubov, M. M. Firsova, and T. M. Molokova.
     Temperature dependence of the dielectric constant in crystalline
     and amorphous quartz. Kristallografiya, Vol. 8, No. 1,
     1963.

254. A. T. Bondarenko. Investigation of the temperature dependence of dielectric constant and loss in rocks at various frequencies. Izv. Akad. Nauk SSSR, Ser. Geofiz., No. 3, 1963.

255. B. I. Vorozhtsov. Some electrical properties of amorphous quartz at high frequency, high voltage and high temperature. Izv. Tomsk. Politekhn. Inst., Vol. 91, 1956.

256. V. A. Ioffe. Dielectric loss in silicate glass. Zh. Tekhn. Fiz., Vol. 24, No. 4, 1954.

257. A. S. Sonin and I. S. Zheludev. Dielectric properties of mono-crystalline boracite. Kristallografiya, Vol. 8, No. 2, 1963.

258. V. A. Ioffe and G. I. Khvostenko. Anomalous dispersion of dielectric constant in feldspar. Dokl. Akad. Nauk SSSR, Vol. 118, No. 4, 1958.

259. V. A. Ioffe and I. S. Yanchevskaya. Dielectric loss in feldspar. Zh. Tekhn. Fiz., Vol. 28, No. 10, 1958.

260. V. A. Ioffe and I. S. Yanchevskaya. Dielectric properties of some alumino-silicates. From: Fizika Dielektrikov, Akad. Nauk SSSR, 1960.

261. G. I. Skanavi, Ya. M. Ksendzov, V. A. Trigubenkov, and V. G. Prokhvatilov. Non-signo-electric dielectrics with high dielectric constant. Izv. Akad. Nauk SSSR, Ser. Fiz., No. 3, 1958.

262. Ya. M. Ksendzov. Effect of impurities on dielectric properties of rutile. Izv. Akad. Nauk SSSR, Ser. Fiz., No. 3, 1958.

263. R. L. Smith-Rose. The electrical properties of soil for alternating currents of radio frequencies. Proc. Roy. Soc., Ser. A, Vol. 140, No. 841, 1933.

264. M. K. Chakravarty and S. R. Khastigir. Direct determination of the electrical constant of soil at ultra-high radio-frequency. Phil. Mag. and J. Sci., London, Vol. 25, No. 170, 1938.

265. I. E. Balygin and V. I. Vorob'ev. Measurement of the dielectric constant and conductivity of soil. Zh. Tekhn. Fiz., Vol. 4, No. 10, 1934.

266. A. P. Kraev. Geoelectric Fundamentals. Gostekhizdat, 1951.

267. A. V. Veshev. Laboratory investigation of the relationship of dielectric constant ε and conductivity σ of rock samples

to the frequency of an electromagnetic field. In: Geofizi-
cheskie Metody Razvedki, Gosgeoltekhizdat, 1955.

268. A. G. Ivanov. On the relationship of the active resistivity
of a rock to the frequency of the current. Vopr. Teorii i
Praktiki Elektrometrii, 1961.

269. B. I. Korennov and G. M. Chernyi. Laboratory investigation
of the dispersion of the dielectric constant in rock samples.
Geol. i Geofiz., No. 11, 1962.

270. G. V. Keller and P. H. Licastro. Dielectric constant and
electric resistivity of natural state cores. Geol. Survey
Bull., 1052-H, 1959.

271. A. N. Efremov. On the anomalous dispersion found in
some dielectrics in the audio frequency range. In: Fizika
Dielektrikov, Akad. Nauk SSSR, 1960.

272. S. Dokoupil and J. Koppinsky. The attenuation of electromagnetic waves in rocks. Inst. Radio Engin. and Electron-
ica., Cescosl. Acad. Sci., Prague, 1962.

273. J. Miyasita and K. Higasi. Dielectric investigation on coals.
Bull. Chem. Soc. Japan, Vol. 30, No. 1, 1957.

274. J. N. Das. The dielectric and piezoelectric behavior of
pyrolusite. Z. Phys., Vol. 155, No. 4, 1959.

275. C. G. Koops. On the dispersion of resistivity and dielectric
constant of some semi-conductors at radio frequencies.
Phys. Rev., Vol. 83, No. 1, 1951.

276. Yu. L. Sevast'yanov, N. N. Dolgopolov, Yu. L. Bur'yan,
and V. S. Margolin. Measurement of the electrical characteristics in warm and dielectrically heated coal. Dokl.
Akad. Nauk SSSR, Vol. 74, No. 4, 1950.

SUPPLEMENTARY GUIDE

by George V. Keller

Supplementary Guide to the Literature on Electrical Properties of Rocks and Minerals

Introduction

The monograph by Parkhomenko is the most extensive discussion of the electrical properties of rocks which has appeared to date. While reference is made to a number of non-Russian papers, the work does not represent as comprehensive a coverage of research in the United States and other countries as it might. The recently published Handbook of Physical Constants [21] serves to fill this gap, but this reference, too, fails to cover the most recent researches. The following pages contain an extensive though not exhaustive guide to the English-language literature on various facets of research on the electrical properties of rocks and minerals.

Basic Principles

If one considers the electrical properties of rocks to be those physical properties which affect the distribution of currents or the propagation of electric fields in rocks, then there are three essentially independent, basic properties and a host of secondary properties. In order to specify how current will flow in a rock, it is necessary to know the electrical resistivity (or conductivity), the dielectric constant, and the magnetic permeability. Usually, magnetic permeability is considered to be a magnetic property rather than an electrical property, therefore, no discussion of magnetic permeability is contained in Parkhomenko's monograph. Rather, she discusses resistivity, dielectric constant, and dielectric loss, the last being a combination of the first two properties.

There are a number of basic non-Russian references dealing generally with the electrical properties of materials which can be listed to supplement the bibliography in Parkhomenko's monograph. A basic text on solid state physics which covers the subjects of metallic and semiconductor conduction mechanisms is one by Kittel [78], or one by Holden [55]. Pauling's [103] text on chemical bonding is a useful supplement in extending these principles to specific materials. More advanced and comprehensive discussions of the theory of conduction in metals are contained in a text by Mott and Jones [96] and in tutorial papers by Slater [118] and by Gerritsen [43]. Garlick [42] has published a paper on photoconductivity, a subject discussed in some detail by Parkhomenko. Electronic semiconduction is described in a paper by Madelung [84].

In addition to the very thorough treatment of electrolytic conduction in solids in the work of Ioffe [59] and of Frenkel [34], alternate mechanisms for the generation of charge carriers in an ion-conducting solid are discussed by Lidiard [82]. Dielectric loss in glass is discussed by Stevels [120].

In most rocks at the earth's surface, conduction through electrolytes filling the pores of a rock is more important than conduction through solid minerals grains. Conduction in liquid electrolytes is covered in many texts on physical chemistry, such as that by Glasstone [45], though a compact discussion of the subject has been given by Darmois [26].

There are a number of reference works in English on the theory of dielectric polarization and loss. von Hippel's [128] book is an excellent introduction to the subject of polarization, while more advanced discussions of the theories of polarization are given by Böttcher [11] and by Frölich [41]. Other specialized papers dealing with various aspects of polarization phenomena are those by Popper [107] on the properties of non-oxide ceramics, by Sutton [124] on the dielectric properties of glass, by Plessner and West [106] on the properties of high-permittivity ceramics, by Brown [15] on the properties of artificial dielectrics, and by Hasted [49] on the dielectric properties of water. Review articles on mechanisms of dielectric polarization and loss include those by Cole [22], Meakins [90], and Wyllie [137].

These references do not deal specifically with the electrical properties of rocks and minerals, but, rather, provide the theoretical background for explaining the observed electrical properties of materials. Much of the literature dealing with the electrical properties of rocks and minerals is descriptive in nature rather than explanatory, as may be seen from the titles in the bibliography at the end of this supplement.

Electrical Properties of Rocks Containing Conductive Minerals

Rocks at the earths surface normally conduct electricity because they contain water within their pores. A few rocks contain large enough concentrations of conductive minerals such as the various sulfides, magnetite, graphite, or carbon to render them conductive without any contribution from water filling the pore spaces. Although such rocks might be considered to be volumetrically insignificant, their value as ores for various metals makes them quite important.

The earliest recognition of the fact that economic ores might be located on the basis of their electrical conductivity appears to have been by Barus [9] in 1882. There were a number of early measurements of the comparative abilities of various ore minerals for carrying current, such as the work of Backström [8], which was published in 1888, but the first comprehensive study of the electrical properties of such minerals using modern techniques was the work of Harvey [48], published in 1928. More recently an extensive study of the conductivity of minerals has been published by Parasnis [101].

The data which are available on the conductivity of conductive minerals indicate that the value of conductivity for a specific mineral cannot be precisely specified. Variability in resistivity is to be expected, inasmuch as the electrical properties are determined primarily by small amounts of impurities and by imperfections, rather than by the bulk composition of the material. Harvey also points out that in many cases erroneously high values of resistivity may be measured on samples which have hairline cracks. In such a case, the lower values of resistivity reported for a mineral are probably more correct than the higher values.

In view of the variability of resistivity in conducting minerals, it is necessary to make a large number of measurements and treat them statistically, as is done in the probability distribution curves

in Fig. 24 of Parkhomenko's monograph. Some minerals show a
relatively narrow range of values reported for resistivity, and
these narrow ranges should be viewed with suspicion, inasmuch
as the lack of scatter is more likely due to a limited number of
measurements than to a uniformity of character for the mineral.

The probability distribution curves in Parhomenko's Fig. 24
exhibit patterns which reflect the factors contributing to the variability of resistivity values measured on a single sample. Three
types of distribution curve may be recognized:

1. A distribution curve with a well defined maximum value,
but with the distribution skewed in the direction of high resistivities, rather than being symmetric about the maximum, such as
the curve for bornite. Bornite is a relatively simple mineral with
a well-specified chemical formula. The high values of resistivity
contributing to the skewing of the distribution probably represent
experimental error (which will tend usually to be in the direction
of high values of resistivity in a highly conductive material) and
the effects of impurities (which would also be more likely to increase the resistivity than to decrease it).

2. A distribution curve with several well-defined maximums,
as shown by the distribution curve for magnetite. This suggests
that the mineral has several different types of internal structure,
insofar as electrical conduction is concerned. In contrast with
bornite, magnetite does not have a well-defined chemical formula,
but is a combination of ferrous and ferric oxides. The several
peaks in the distribution curves probably correlate with preferred
modes of mixture of these two chemical compounds.

3. A distribution curve with a single maximum, but skewed
towards lower resistivities, rather than being symmetric about
the maximum. An example is the curve for galena. This type of
behavior is observed with minerals which are normally more resistant than the majority of ore minerals. In such minerals, conduction is controlled by the addition of the proper impurities. The
presence of these impurities causes some samples to have a considerably lower resistivity than the purer forms of the same chemical compound.

Consideration of such distribution curves leads to the conclusion that the value of electrical resistivity is not a diagnostic

property useful in identifying specific ore minerals. This lack of diagnosticity is unfortunate, but it is of no great importance in the search for metal ores. Commonly, ore bodies contain a wide variety of sulfide minerals, and the most abundant of the minerals in the ore body may not be the one which makes the ore body economic. In considering the resistivity of an ore deposit, we need be concerned only with the total complex of conductive minerals in estimating how resistivity may correlate with ore content.

Although there are a great number of minerals which are excellent conductors, only a few of them occur in large enough amounts to render a rock conducting. These include magnetite, specular hematite, carbon, graphite, pyrite, and pyrrhotite. The habit of a mineral (the geometric form in which it habitually occurs) is as important as the amount in determining the effect on the bulk resistivity. Most of the minerals which are known to render rocks conductive occur in dendritic patterns, so that only a small amount of the mineral is required to decrease the resistivity of a rock tremendously. Other conducting minerals occur only as isolated grains, or they may be enveloped in a coating of an insulating gangue mineral such as calcite or quartz.

There has been relatively little work reported on the correlation between resistivity of a rock and the amount of conducting minerals in the rock, other than a group of papers from the U. S. Geological Survey [1, 68, 143, 144, 145, 147, and 148]. An example of how the amount of conducting mineral affects the overall conductivity of a rock is shown by the data in Fig. 1. In this example, the rock is pyrrhotite-bearing gabbro from Maine [1]. Each plotted point represents the average of a large number of measurements; otherwise, the variability of individual measurements would mask the character of the correlation between resistivity and ore grade. It is apparent that there is a threshhold content of about 5% pyrrhotite, below which the presence of the pyrrhotite has little or no effect on the overall conductivity and above which the presence of the pyrrhotite renders the rock highly conductive. Zablocki [147] points out that this threshhold content for conducting minerals is highly sensitive to the texture of these minerals, even when only one species of conducting mineral is being considered. For example, in the Lake Superior district, 8% magnetite content is enough to reduce the resistivity of chert from several thousand ohm-meters to less than one ohm-meter, primarily

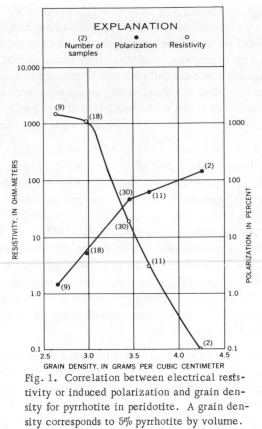

Fig. 1. Correlation between electrical resistivity or induced polarization and grain density for pyrrhotite in peridotite. A grain density corresponds to 5% pyrrhotite by volume.
From Anderson [1].

because the magnetite grains are overgrown by needle-like crystals of hematite, which form a conducting network. On the other
hand, in a replacement-type magnetite deposit in California where
there are no such hematite overgrowths, 25% magnetite by volume
is required to reach the threshhold of conduction.

Graphite (or carbon) and pyrite are probably the minerals
which most commonly cause a rock to be conductive. Carbon,
graphite, and pyrite may be abundant as accessory minerals in
slates, but it is uncertain which of these minerals contributes to
the conduction observed in such rocks. It has been observed that
"stratigraphic units" within slate may be rendered conductive along
the strike of a slate outcrop for miles, so that they may be mapped
with conductivity measurements [40].

Electrical Properties of Water-Bearing Rocks

The electrical resistivity of a water-bearing rock will depend on the amount of water present, the salinity of this water,
and the way in which the water is distributed through the rock.

The relationship between rock resistivity and water content
has been of interest primarily to petroleum explorationists, inasmuch as resistivity measurements serve as a basis for estimating
fluid saturations in potentially oil-bearing rocks. As a result, the
great bulk of research done on resistivity—water content relationships has been done with the rock most likely to be oil-bearing—
porous, marine sedimentary rocks. Very little work on non-sedimentary rocks has been reported.

Measurements applicable to oil exploration have been reviewed by Pirson [105]. A general discussion of the characteristics of ground water has been given by Meinzer [91].

In view of the fact that appreciable amounts of current flow
through the water phase in a water-bearing rock, one must be
concerned with the electrical properties of naturally occuring
waters. No mention is made of the electrical properties of water
in Parkhomenko's monograph, but there is a great volume of information available, both on the composition of dissolved solids in
natural waters and on the electrical resistivity. The significance
of chemical analyses of dissolved solids in natural waters is discussed in a definitive manner by Hem [52] in a monograph which
also contains several hundred references to other literature.
Methods of comparing water analyses are also described by Sage
[113].

Typical of reports on the properties of ground water in restricted areas are papers by Crawford [24], Case [17], Ayers et
al. [7], Miles [93], Puzin [109], and Stringfield et al. [121]. Hundreds of such papers have been published, particularly in the
series of Water Supply Papers published by the U. S. Geological
Survey, though these tend to emphasize surface waters rather than
subsurface waters.

The resistivity of a water sample may be calculated from a
chemical analysis of the dissolved solids on the basis of the mobilities and activities of the ion species in solution. In practice, because some ions are much more common than others in natural

waters, a semiempirical technique is used to express an actual water analysis which lists many salts in solution in terms of an equivalent content of a single salt, usually sodium chloride [29, 94, 95]. This is done by using weighting factors to replace the effect of the less common salts with an equivalent amount of the abundant salt. It is then a simple matter to determine the resistivity of the equivalent sodium chloride solution from standard tables, such as those in the International Critical Tables [58].

Ground water may be classified either on the basis of origin or on the basis of chemistry. Water which is believed to be a fossil remanant of the water in which a sedimentary rock was originally deposited is termed connate. The original water contained in igneous rocks is termed juvenile. Water introduced recently into rocks from the atmosphere is termed meteoric. Usually, any of these waters have been so altered by absorption, evaporation, dilution, and other processes that they are not similar chemically to the water originally in the rock.

Depositional environment does control the properties of ground water to some extent; marine sedimentary rocks generally have a higher concentration of salt in solution in their pore waters than do continental sedimentary rocks. Other effects are also important in determining the salinity of ground water; Paleozoic rocks generally are saturated with saltier water than more recent rocks, the higher salinity being a result of the loss of water on compaction and the addition of salts by solution of mineral grains. The salinity of water in fine-grained rocks tends to be higher than the salinity in coarse-grained rocks because the greater adsorption capacity of rocks with large internal surface area holds salts from diffusing.

A geochemical classification of waters which supposedly catalogs waters according to depositional environment recognizes (1) bicarbonate waters, (2) sulfate waters, and (3) chloride waters. Chloride waters are characteristic of sedimentary rocks deposited in a marine environment. Sulfate and bicarbonate waters appear to be juvenile waters in igneous rocks or normal waters in continental sediments. These three classes of water differ widely in normal salinity — bicarbonate waters have a median salinity of only about 400 parts per million, sulfate waters have a median salinity of about 5000 parts per million and chloride waters have salinities in excess of 10,000 parts per million. For comparison, present day ocean water is of the chloride type and has a total salinity of 33,000 to 35,000 parts per million.

Table 1.   Average Resistivities of Ground Water
(from Chebotarev [19])

Rock from which water samples were taken	Number of samples	Resistivity at 20°C (Ω-m)	
		median	range
Igneous rocks Europe	314	7.6	3.0 − 40
Igneous rocks, South Africa	175	11.0	0.5 − 80
Metamorphic rocks, South Africa	88	7.6	0.86 − 80
Metamorphic rocks, Precambrian of Australia	31	3.6	1.5 − 8.6
Recent − Pleistocene continential sediments, Europe	610	3.9	1.0 − 27
Recent − Pleistocene sediments, Australia	323	3.2	0.38 − 80
Tertiary sedimentary rock, Europe	993	1.4	0.7 − 3.5
Tertiary sedimentary rock, Australia	240	3.2	1.35 − 10
Mesozoic sedimentary rock, Europe	105	2.5	0.31 − 47
Paleozoic sedimentary rock, Europe	161	0.93	0.29 − 7.1
Chloride waters from oil fields	967	0.16	0.049 − 0.95
Sulfate waters from oil fields	256	1.2	0.43 − 5.0
Bicarbonate waters from oil fields	630	0.98	0.24 − 10

Chebotarev [19] has given an excellent summary of a great number of ground water analyses, and a few of his average values are listed in Table 1.

Many thousands of water analyses from oil wells have been collected by oil companies, but most of these data have not been published. Recently, the Society for Professional Well Log Analysts has sponsored the publication of catalogs of connate water resistivities by its member sections. One such catalog is currently available [27] and others are anticipated.

The resistivity of the water contained in a rock cannot be determined solely from the properties of that same water outside the rock. When water is distributed through a pore structure, there is an interaction between the water and the rock which may alter the conductivity of the water appreciably. Several types of interaction may take place, either increasing or decreasing the conductivity of the water in the pore structure. Water is adsorbed to the surfaces of most mineral grains and, as a result, the viscosity of the water in place in a rock is greater than the viscosity of the water in bulk. This increase in viscosity reduces the mobility of ions and increases the bulk resistivity of the rock [65]. Usually, this increase in resistivity is offset by a decrease in resistivity associated with partial solution of the rock by the water in the pore structure. Clay minerals with exchangeable ions participate in this partial solution [54, 89, 102, 108, 104, 114, and 133]. Exchangeable ions contained in the clay minerals tend to go into solution when the clay minerals come in contact with water, forming an ion halo about each clay particle. These ions contribute to conduction with reduced mobilities and return to the clay particle when the water is removed from the rock. Thus, they have no effect on the salinity measured on a sample of water removed from a rock. This added conductivity is most important in rocks saturated with dilute water and is insignificant in rocks saturated with saline water. One rarely finds that the resistivity of water in place in a rock exceeds $10 \ \Omega\text{-m}$, no matter how pure the water may be when extracted from a rock.

The major features of the relationship between water resistivity in bulk and in place in a pore structure is shown in Fig. 2 [65]. The interaction effects are most significant in fine-grained rocks, least significant in coarse-grained rocks.

The amount of water in a rock, as well as its salinity, is important in determining the bulk resistivity. The maximum amount of water which can be held by a rock is limited by the porosity − the fraction of the rock volume not occupied by solids. In some cases, the water content may be less than the pore volume available if the pore space is partially occupied by hydrocarbons, as in oil fields, or by atmospheric gases, as in near-surface rocks.

As in the case of water resistivities, porosities have been most thoroughly studied for marine sedimentary rocks. A sum-

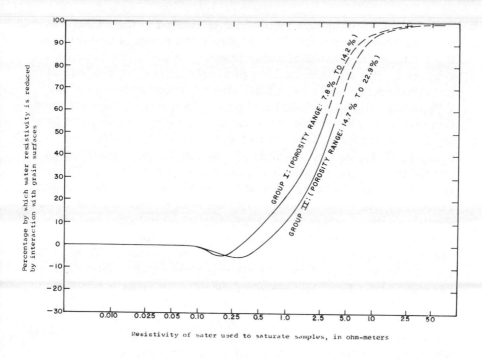

Fig. 2. Changes in resistivity of water in a Jurassic sandstone caused by interaction with the grain surfaces. Measurements were made on 412 samples divided in two groups, one with high porosity and the other with low porosity. From Keller [65].

mary of such porosity determinations has been given recently in the Handbook of Physical Constants [25], and an extensive bibliography is included in an earlier paper by Manger [85].

It is well known that porosity may vary widely, from essentially zero in dense rocks to nearly 100% of the total rock volume in some volcanic rocks and some corals and uncompacted muds. Usually, dense rocks at the earth's surface and to a depth of at least one kilometer contain appreciable porosity in the form of joints, accounting for 0.2 to 0.3% of the rock volume. At greater depths, pores may be closed by overburden pressure, but there is no direct evidence to indicate what the least porosity in the lower part of the earth's crust might be.

In determining how resistivity is related to the water content of a rock, it is difficult to separate the effect of texture from the effect of the amount of water present. One might compute the resistivity of a uniform packing of simple geometric shapes, but such models appear to have little resemblance to actual rocks. Computation of the resistivity of a rock in terms of the water content and textural parameters might be possible, as indicated by Wyllie and Gregory [139], Wyllie and Rose [140], Wyllie and Spangler [141], and by Towle [126], but the textural parameters needed in such computations are not easily measurable. These difficulties have led to the use of empirical correlations between resistivity and water content. The earliest of such studies apparently are those of Sundberg [122], published in 1932, and of Jakosky and Hopper [62], published in 1937. However, such relationships are today known as Archie's law, because Archie [3-5] was the first to recognize that the graphical relationship could be expressed by a fairly simple formula

$$\rho_t = \rho_w \phi^{-n}$$

(1)

where ρ_t is the bulk resistivity of a rock, ρ_w is the resistivity of the water as it actually exists in the pore space, considering the interaction effects discussed above, and ϕ is the porosity, assumed to be water-filled, expressed as a volume fraction. The exponent n is an expirically derived parameter characteristic of the texture of the rock. The value varies from about 1.3 in loosely packed granular material to about 2.2 in well-cemented granular rocks [20, 138].

Some investigators believe that the relationship is better expressed by introducing a second empirical parameter, a, which multiplies the porosity term [105]

$$\rho_t = a\rho_w \phi^{-n}$$

(2)

The value for the parameter n varies from 0.6 to 1.3 in marine sedimentary rocks, and appears to be related to the texture of the rock.

A few data are available for non-sedimentary rocks [67, 75, 77] which indicate that the empirical equation (1) or (2) holds

Table 2. Archie's Law Expressions for Various Rock Types

1. Weakly cemented detrital rocks, such as sand, sandstone and some limestones, with a porosity range from 25 to 45%, usually Tertiary in age

$$\rho_t/\rho_w = 0.88 \; \phi^{-1.37}$$

2. Moderately well-cemented sedimentary rocks, including sandstones and limestones, with a porosity range from 18 to 35%, usually Mesozoic in age

$$\rho_t/\rho_w = 0.62 \; \phi^{-1.72}$$

3. Well-cemented sedimentary rocks with a porosity range from 5 to 25%, usually Paleozoic in age

$$\rho_t/\rho_w = 0.62 \; \phi^{-1.93}$$

4. Highly porous volcanic rocks, such as tuff, aa and pahoehoe, with porosity in the range from 20 to 80%

$$\rho_t/\rho_w = 3.5 \; \phi^{-1.44}$$

5. Rocks with less than 4% porosity, including dense igneous rocks and metamorphosed sedimentary rocks

$$\rho_t/\rho_w = 1.4 \; \phi^{-1.58}$$

reasonably well with the proper assignment of the empirical parameters. Typical expressions in the form of Archie's law for various types of rocks are listed in Table 2.

The expressions in Table 2 apply only when a rock is completely saturated with water, and some consideration must be given to the possibility that the water may be partially displaced by a nonconducting fluid. It has been observed that the resistivity of a rock increases in proportion to the inverse square of the fraction of the pore space filled with water, provided the water which remains coats the grains uniformly [64]

$$\rho_t = a\rho_w\phi^{-m}/S^2 \tag{3}$$

where S is the fraction of the pore space filled with water having the resistivity ρ_w.

If the water saturation drops below the critical level required to coat the grains (an amount which ranges from 0.2 in sandstones to 0.8 in dense, jointed rocks), the resistivity will be much higher than that given by equation (3). Also, if an insulating liquid such as oil preferentially wets the grain surfaces, the resistivity of a rock may be as much as ten times greater than if the same amount of water wetted the mineral surfaces [64, 81, 125].

Of the three primary rock characteristics which affect resistivity — the amount of contained water, the salinity, and the manner of distribution — all are not equally important. The resistivity of rocks varies most widely because of differences in water content, which may vary by a ratio of 1000:1 between various rock types at the earth's surface. Because the water content enters Archie's law as a squared quantity, this range in variation of water content contributes a 1,000,000:1 variation in bulk resistivity, other factors being the same. The salinity of the contained water is second in importance, partly because it enters Archie's law only as the first power, and partly because the range of variability for water resistivity is limited by interactions with the minerals in a rock to a ratio of about 500:1. All other factors being equal, variations in water salinity can cause only about a 500:1 range in bulk resistivity. Distribution of the water in the pore structure is the least important factor, inasmuch as it causes no more than a 20:1 range in resistivity, all other factor being equal.

The resistivity of a water-bearing rock is affected by temperature and pressure variations, though to lesser extents than it is affected by the factors discussed above. Unless the temperature range is extreme, the effect of temperature changes on the bulk resistivity of a rock is no different than the effect of temperature on the conductivity of the electrolyte contained in the rock. For most electrolytes, the resistivity depends on temperature as

$$\rho(t) = \rho_{18}/[1 + 0.022(t - 18)] \qquad (4)$$

where $\rho(t)$ is the ambient temperature, ρ_{18} is the resistivity at a reference temperature of 18°C, and t is the temperature.

Temperatures within the first few miles of the earth's surface rise gradually with depth at a rate of about 0.5°C per hundred
feet [80] in sedimentary rocks and about 0.2°C per hundred feet in
igneous rocks. This increase in temperature causes an increase
in the conductivity of the rock, at least until the boiling point of
water is reached. However, the boiling point also increases with
increasing pressure, and where the thermal gradient in the earth
is normal, the boiling point is not exceeded at any depth. Only in
rare cases is the temperature in the earth high enough to remove
water from the rock as steam. This may occur in regions of recent volcanism, such as Hawaii, where near-surface pools of
molten rock raise the temperature of the surrounding rock above
the boiling point of water [39]. In such cases, the conductivity of
the rock is that of the solid minerals.

The behavior of resistivity in water-bearing rocks below the
freezing point is important because about one-seventh of the earth's
surface is frozen or permanently covered with ice or snow. One
might expect freezing to increase the resistivity of a rock quite
markedly, but experiments indicate that such an increase, while
observed, is not as large as one might expect. The low frequency
resistivity of ice is very high [119], being about 10^7 Ω-m, so ice
should contribute no more to conduction in a rock than many of the
common rock-forming minerals.

Measurements of rock resistivity at sub-freezing temperatures have shown that at -12°C, the resistivity is only about 10 to
100 times larger than the resistivity measured at +18°C [28]. The
behavior of resistivity in granular rocks as the temperature is
lowered below the freezing point is shown in Fig. 3.

The fact that freezing has only a moderate effect on resistivity may be explained as follows:

1. Most ground water is moderately saline, and the presence of salt in solution lowers the freezing point of an electrolyte.
Moreover, when an electrolyte is frozen slowly, salt ions migrate
from the solidifying phase to the still-liquid phase, increasing the
salinity of this fraction and further lowering the freezing point.
As a result, freezing can be considered as merely a reduction in
the fraction of pore space occupied by water for temperatures
down to -60°C.

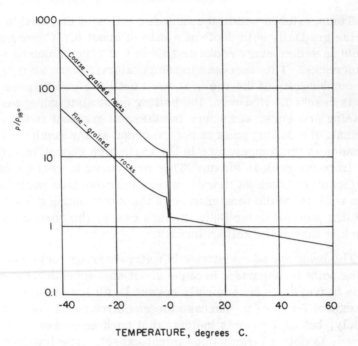

Fig. 3. Variation of resistivity of water-bearing rocks above and below
the freezing point. From Dumas [28].

2. Increased pressure also lowers the freezing point of
water. Water adsorbed on grain surfaces is under great pressure
and will not reorient into ice crystals until the temperature is
considerably below the normal freezing point.

These two factors, pressure and salinity, cause freezing in
rocks to take place over an extended temperature range rather
than at a single temperature. The greater the salinity of the pore
water, the lower will be the temperature at which freezing first
starts, and the finer the grain size, the broader will be the temperature range through which freezing continues.

The effect of pressure on the resistivity of a water-bearing
rock at normal temperatures has been studied by a number of investigators, with essentially the same results as summarized by
Parkhomenko for Russian investigations [6, 31, 44, 51, 53, 56,
110, and 136]. A review of these studies on the effect of pressure
on the resistivity of water-bearing rocks along with an extensive
bibliography is given by Helander and Campbell [50].

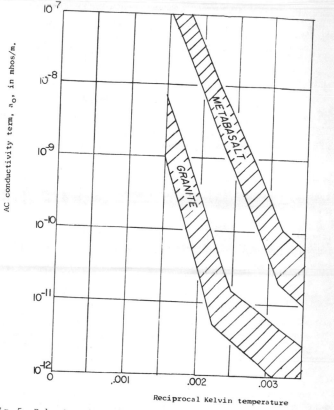

Fig. 5. Behavior of the AC conduction term as a function of temperature and lithology.

Relatively few data have been reported for measurements of conductivity of dry rocks as a function of frequency and temperature [70, 76, 146], although very extensive measurements are being made at the U. S. Geological Survey [117]. A typical set of data for the relationship between resistivity and frequency, measured at various temperatures, is shown in Fig. 4.

AC conduction appears to be of two types — one which is dominant at low temperatures and which is strongly frequency dependent, and one which is dominant at high temperature and which is independent of frequency. At low temperatures, the observed conductivity increases with frequency in nearly direct proportion, while at high temperatures, the conductivity remains nearly constant over the frequency range from DC to several megacycles per second.

Electrical Properties of Dry Rocks

Deep within the earth, it appears that solid conduction through mineral grains may become more important than conduction through water filled pore spaces. The reason for this is twofold— at great depths, pore structures are compressed or closed by overburden pressure so that there is less free water available than in near-surface rocks, and the temperature increases with depth, markedly reducing the resistivity of the minerals comprising the rocks. There has been considerable interest recently in measuring the electrical properties of the crust at depths ranging from a few kilometers to tens of kilometers [70]. On the basis of such studies, it appears that the earth's crust and upper mantle may be divided grossly into three regions – a surface region in which the rocks contain significant amounts of water and are therefore moderate to good conductors; an intermediate zone in which the rock contains very little water, but is not hot enough to have appreciable mineral conductivity so that the rock has a very high resistivity; and a third, lower zone starting at depths of the order of 10 to 20 km where the temperature is high enough to cause significant conduction. In the surface zone, the resistivity is determined by the presence of moisture in the rocks, with the resistivity usually falling within the range from 1 to 10,000 Ω-m. In the intermediate zone, no well-substantiated resistivity determinations have been made, but the resistivity probably falls in the range from 50,000 to 500,000 Ω-m. At greater depths, down to 60 km, the resistivity is of the order of 10 to 100 Ω-m. Discussions of the determination of electrical properties deep in the crust are found in papers by Eckhardt et al. [30], by Keller [74], by Lahiri and Price [79], by McDonald [88], by Migeaux et al. [92], and by Rikitake [111].

Much of the information we have about the probable electrical properties deep within the earth has been obtained as a result of laboratory studies of the electrical properties of rocks and minerals under simulated conditions of temperature and pressure, similar to the studies reported by Parkhomenko in her monograph. It has been found that changes in temperature evoke much larger changes in the resistivity of a dry rock than do changes in pressure, therefore the majority of laboratory investigations has dealt with the electrical properties of samples as a function of temperature only. Also, most of the work has been done with direct current, and very few studies of the frequency dependence of resistivity have been published as yet.

　　　Early work on electrical properties under conditions simulating the interior of the earth have been reviewed by Tozer [127], who concluded on the basis of theoretical considerations that the conductivity of the mantle beyond a few hundred kilometers depth is probably electronic in nature. Because the mantle is believed to be comprised of basic or ultrabasic minerals, most of the laboratory investigations have dealt with such minerals [13, 14, 47, 57, 60, and 112]. In such minerals, it is difficult to determine whether electron or ion conduction is dominant under various experimental conditions, which have involved temperatures to 1200°K and pressure to 42 kb.

　　　Other investigators, including Noritomi [97-99], Noritomi and Asada [100], and Coster [23], have been more concerned with rocks than with individual minerals. Noritomi has collected electrolysis products after direct-current conduction tests, and concludes that ion transfer is the dominant mechanism of conduction. Zablocki [146], making measurements with serpentinite samples containing significant amount of water of hydration, found that the presence of such water rendered the rocks about 100 times more conductive than the same rocks with the moisture removed.

　　　In the studies of both rocks and minerals, several regions have been recognized in the conductivity–temperature relationship. Within each of these regions, the logarithm of conductivity is roughly proportional to the inverse absolute temperature. Because the linear relationship is not exact, the curves may be broken into a number of segments, but considering the approximation involved, it seems reasonable to restrict such an approximation to two segments only, described with the equation

$$\sigma = A_1 e^{-U_1/kT} + A_2 e^{-U_2/kT}, \tag{5}$$

where σ is the conductivity in mhos per meter, the constant A_1 and A_2 are dependent on the numbers of charge carriers available for conduction and their mobilities, and U_1 and U_2 are the activation energies for ions if conduction is ionic, or twice the activation energies if conduction in electronic. Typical values for these parameters are listed in Table 3.

Table 3. Values for the Parameters Used in Describing the Relationship Between Conductivity and Temperature in Dry Rocks

Material	Low-temperature parameters		High-temperature parameters		Transition temperature (°K)
	A_1 (mho/m)	U_1 (eV)	A_2 (mho/m)	U_2 (eV)	
Acidic igneous rocks (granite granodiorite)	$3 \cdot 10^{-3}$ to $5 \cdot 10^{-4}$	$0.5 - 0.6$	$10^4 - 10^6$	$2.3 - 2.7$	$800 - 1000$
Basic igneous rocks (gabbro, peridotite)	0.005 to 0.01	$0.7 - 0.8$	$10^6 - 10^6$	$2.0 - 2.3$	$700 - 800$
Hydrous igneous rocks (serpentenite)	0.01 to 0.1	$0.3 - 0.5$	$10^4 - 10^5$	$1.5 - 2.0$	$600 - 700$

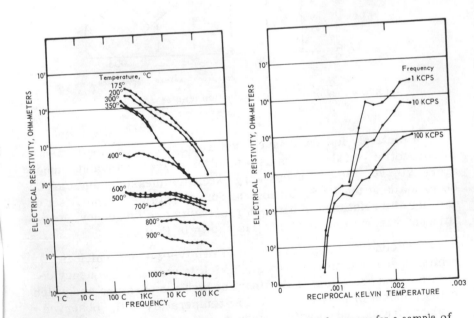

Fig. 4. Relationship between resistivity, temperature, and frequency for a sample of coarse-grained gabbro.

The total conduction can be written as the sum of two terms, an AC conductivity and a DC conductivity

$$\sigma_{total} = \sigma_{AC} + \sigma_{DC} \qquad (6)$$

The DC conductivity term is the same as the conductivity given in equation (5). Little is known about the AC term, but a few data suggest that it varies with frequency according to the law

$$\sigma_{AC} = A(T)\omega^{n(T)} \qquad (7)$$

where ω is frequency in radians per second and $a(T)$ and $n(T)$ are functions of Kelvin temperature.

The parameter $a(T)$ exhibits the usual characteristics of a thermally activated conductivity. When it is plotted as a function of temperature (Fig. 5), it is apparent that $a(T)$ can be filled with the following equation over a limited temperature range:

$$a(T) = a_0 \, e^{-T_c/T} \qquad (8)$$

where $T_c = W/k$ represents an excitation energy and a_0 is a base level for the function $a(T)$. The curves consist of two segments, with transition from low-temperature to high-temperature behavior taking place at about 100 to 150°C. Values for T_c and a_0 are listed in Table 4 for granite, metabasalt, and rhyolite, the only three rocks for which data are available.

The value for $n(T)$ in equation 7 also shows a systematic variation with temperature and rock type. The value is closer to unity at low temperatures and for granitic rocks than at high tem-

Table 4. Parameters Describing AC conducting
in Dry Rocks

Rock type	a_0, Ω^{-1}/m	T_c, °K
Metabasalt	$3.7 \cdot 10^{-4}$ to $1.35 \cdot 10^{-3}$	$5140 - 5240$
Granite	$2.95 \cdot 10^{-5}$ to $1.5 \cdot 10^{-4}$	$6400 - 6960$
Rhyolite	0.61	$13,500$

peratures or for basaltic rocks. The behavior of the exponent is
shown graphically in Fig. 6.

    There has been virtually no geophysical interest in the elec-
trical properties of rocks in the molten state, except for some
work by Frischknecht and Anderson [39] and by Keller et al. [77]
in measuring the resistivity of molten lava at the Kilauea Caldera
in Hawaii. A value of 0.2 to 0.3 Ω-m was observed for rocks a
few tens of degrees above their melting points, close enough to
the surface to be under negligible pressure. This is lower by a
factor of 10 to 20 than the resistivity of similar basaltic rocks a
few tens of degrees below their melting point. The electrical
properties of molten rocks have been studied more extensively by
geochemists (see, for example, Bloom [10]).

Dielectric Constants, Dielectric Loss, and Induced Polarization

    In her monograph, Parkhomenko chooses to treat dielectric
polarization and dielectric loss as separate properties and to exclude the subject of induced polarization, leaving it for a later
publication. This division of the various aspects of electric polarization in rocks is somewhat artificial, and a simultaneous
treatment of all three topics is helpful in understanding the mechanisms involved.

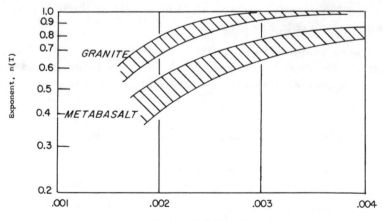

Fig. 6. Variation of the exponent n(T) in the AC conduction term with
temperature for granite and metabasalt.

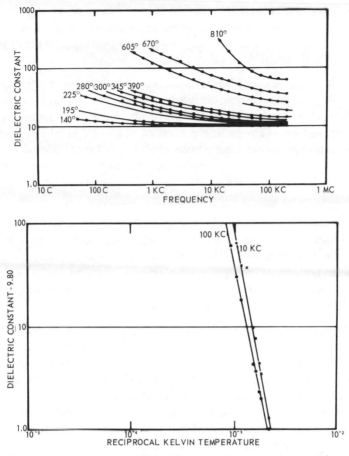

Fig. 7. Relation between dielectric constant, temperature, and frequency for a sample of rhyolite tuff.

Polarization consists of the separation of charge in a material when an electric field is applied. The dielectric constant is
a measure of the relative abilities of materials to store charge for
a given applied field strength, while dielectric loss is a measure
of the relative amounts of charge transferred in conduction and
stored in polarization. In simple materials, separation of charge
can be accounted for by the shift of the center of charge in electron
motion relative to the nuclei (electron polarization), the shift of
positively charge nuclei with respect to negatively charge nuclei
(ion polarization), or the rotation of dipolar molecules. The dielectric constants associated with these mechanisms and the di-

electric loss are relatively simple to measure, and extensive
tables of values for rocks and minerals have been given by von
Hippel [129] and by Keller [73], along with bibliographies of papers
describing individual measurements.

Most rocks cannot be considered to be simple materials and,
as a result, the electric polarization which takes place in these
rocks cannot be explained in terms of phenomena at atomic and
molecular scales. There are a host of phenomena which occur in
complex materials that contribute electric polarization which may
be especially intense at low frequencies. Some of these contribute
to what has become known as "induced polarization" in geophysical
exploration.

Basically, induced polarization in the geophysical sense can
be considered to be the extra polarization that occurs in rocks
which cannot be explained in terms of electronic, ionic, or molecular polarization. This polarization can often be explained in
terms of the texture of a rock, and often in terms of electrochemical effects.

Maxwell [87] in 1892 suggested that polarization can be caused
by the peculiar structure of certain types of materials, and such
polarization has come to be known as Maxwell polarization or as
interfacial polarization. If a plate of uniform material is polarized
with an electric field, the magnitude of polarization will be proportional to the intensity of the applied electric field and inversely
proportional to the thickness of the plate. In defining the dielectric
constant for a homogeneous medium, polarization is normalized
for this thickness effect so that the resultant value does not depend
on the dimensions of the sample. However, if a material is made
up of layers of conducting and insulating materials, polarization
occurs only across the insulating layers. Because this layer is
thinner than the plate of the composite material, the polarization
across this layer will be more intense than expected, and the dielectric constant computed for the composite material will be larger
than the dielectric constant for the insulating material alone.

The layers which comprise a composite material need not be
perfect insulators and conductors; interfacial polarization will be
observed whenever a material is made up of layers in which the
product of dielectric constant and resistivity varies from layer to
layer. In fact, the material need not even consist of layers, but

may be any heterogeneous mixture of materials having different
electrical properties. A number of special cases of one material
mixed with another have been analyzed mathematically [35-37, 123].

While interfacial polarization may result in dielectric constants as large as several hundred or several thousand in specially
constructed materials such as ferrites, which are conductive
grains coated with an insulating film, the effect seems to be insignificant in most earth materials in comparison with electrochemical effects.

Electrochemical polarization may take place in any material
in which ions contribute to conduction. Electrochemical polarization is not limited to water-bearing rocks; dry rocks in which
ionic conduction takes place also exhibit such polarization. Typical of this is a set of data for the dielectric constant of a dry
sample of rhyolite measured at various temperatures (Fig. 7).
At high frequencies, the dielectric constant of a rock is approximately the value one might expect on the basis of atomic and
molecular polarizations. At low frequencies, below a few hundred
cycles per second, the dielectric constant at high temperatures
may be ten to twenty times larger then the normal value, with the
major part of the polarization being electrochemical. In wet
rocks, measurements suggest that the electrochemical polarization may be as much as a million times larger than the atomic
and molecular polarizations, but such high values of polarization
are open to question, inasmuch as certain types of experimental
error are difficult to avoid in working with water-bearing materials.

One type of electrochemical polarization occurs when current flows across the interface between an electronic conductor
and an ionic conductor. Such current flow can take place only
when there is oxidation or reduction at the interface. When an
electric field is applied to cause a current to flow across the interface between the two media, that current is limited by the re-
action rate for oxidation or reduction at the interface. The delay
in transfer of current through the interface results in a buildup in
ion concentration of cations or anions in the electrolyte outside
the electronic conductor. This accumulation of ions develops a
Nernst potential, or counter-emf, which tends to accelerate the
rate of oxidation or reduction, until finally there is a uniform rate

of current flow through the material. These ion accumulations, anions on one side of the electronic conductor and cations on the other, represent electric polarization. Considerable experimental and theoretical work has been done on polarization at metal – electrolyte interfaces [16, 18, 46, and 61].

The extra voltage drop required to accelerate the transfer of current across the metal – electrolyte interface is termed overvoltage. The overvoltage is dependent on the current density with which it is measured, with the result that the surface impedance is not linear at the interface. Experimental work indicates a logarithmic relationship between overvoltage and current density, a relationship known as Tafel's law

$$n = a - b \ln j \qquad\qquad (9)$$

where n is the overvoltage, j is the current density at the interface, and a and b are parameters characterizing the electrode surface and the electrolyte with which it is in contact. For metal surfaces in contact with an acid electrolyte, the values of b fall within the limits 0.09 to 0.13, while the parameter a varies over a wide range, from 10^{-5} to 10^{-13} A/cm^2. No data on overvoltage properties of electronically conducting minerals have been published.

The capacity of a metal– electrolyte interface is of the order of 10 to 20 pF/cm^2 at very low frequencies, though widely different values may be observed if the metal surface is rough or tarnished. This capacity leads to high polarizations in rocks containing grains of electronically conducting minerals in contact with electrolytic solutions in the pore spaces, but it is also the source of the most serious type of experimental error involved in measuring the electric properties of water-bearing rocks. The same polarization must occur at the surfaces of the electrodes used in measuring the electrical properties of a sample, and errors from this polarization are extremely hard to avoid, even if a four-terminal electrode system is used.

Overvoltage is not the only source of polarization in water-bearing rocks at low frequencies; membrane polarization is another important source. As described in an earlier section, adsorption of water to grain surface in a fine-grained rock causes an increase in the viscosity of the water and a decrease in the mobility of ions in solution in the water. This variable mobility of ions along the conduction paths leads to accumulations of ions at the ends of pore segments where the mobility is high. The Nernst potentials from these accumulations accelerate the ion movement through the regions of low mobility, until a uniform transfer of current is finally achieved. However, the accumulation of ions at various spots through the pore structure constitutes polarization, just as in the case in which the current flow was impeded by the electronic-conductor electrolyte interface. Other mechanisms may contribute to charge storage in rocks at low frequencies, but these two phenomena appear to be the most important [132].

Both types of polarization have been studied intensively in recent years by mining geophysicists. The most comprehensive publication dealing with low frequency polarization in rocks is that edited by Wait [132]. Overvoltage polarization has been studied by Anderson and Keller [2], while membrane polarization has been discussed by Fraser et al. [33], by Keevil and Ward [63], by Keller [69], by Marshall and Madden [86], and by Schufle [115].

Even with this variety of work, it is difficult to evaluate the results in view of the widely different ways in which induced polarizations are measured. It is possible to compute formally a value for the dielectric constant, with the charge stored by the membrane and overvoltage polarization being considered, but such computations have led to a spirited controversy in the literature concerning the meaning of such values of the dielectric constant. They may be many orders of magnitude larger than the values obtained for atomic and molecular polarization (see Frische and von Buttlar [38] and the discussion of their paper by Wait [131]).

When values are computed for the dielectric constant which include the effects of induced polarization, they are found to vary with frequency for frequencies as low as a fraction of cycle per second. Wagner [130] developed an expression to describe the variation of dielectric constant with frequency for a material made up of a number of relaxation oscillators having a log-normal distribution of time constants. Several investigators [83, 142] have found that the Wagner model is useful in describing the dispersion of the dielectric constant (dispersion is variation with frequency).

Table 5. Values for the Product of Resistivity and low-frequency
Dielectric Constant (Resistivity Expressed in Ω- m, the Dielectric
Constant in F/m)

Rock type	Average value for $\rho\varepsilon$	Range of values	Number of measurements
Hematite ore from Minnesota	10.8	1.5 − 35	14
Glacial till, Minnesota	0.80	0.30 − 3.0	9
Igneous aphanites (rhyolite, basalt)	0.63	0.25 − 2.6	19
Sandstone and siltstone	0.32	0.081 − 0.92	9
Sandstone (Jurassic from Colorado)	0.18	0.015 − 4.0	225
Metamorphosed igneous rocks (granulite, greenstone)	0.17	0.030 − 0.36	10
Acidic igneous rocks (granite, porphyry)	0.16	0.030 − 0.57	8
Limestone and dolomite	0.10	0.030 − 0.25	18
Intermediate igneous rocks (monzonite, diorite)	0.085	0.026 − 0.23	19
Basic igneous rocks (gabbro, chromite, troctolite)	0.0014	0.00033 − 0.033	15

With the Wagner model, the behavior of the dielectric constant
as a function of frequency is described with four parameters: the
dielectric constant at high frequencies, ε_∞; the dielectric constant
at low frequencies or static dielectric constant, ε_0; the principle
relaxation frequency, ω_0; and the standard distribution of relaxation frequencies about this principle frequency, σ. An extensive
table of values for these four parameters has been given by Keller
[66].

Not all of these parameters vary widely in rocks. For example, the parameter σ is nearly constant, with reported values
falling in the range from 0.95 to 1.15. The principle relaxation
frequency varies more widely, ranging from 0.002 to 0.2 rad/sec,
with a median values of about 0.03 rad/sec. The high-frequency value
for dielectric constant, ε_∞, is the value observed for ionic or molecular polarization. The dielectric constant at low frequencies

varies widely, but it appears that the product of ε_0 and the resistivity at low frequencies is characteristic of a particular type of rock. Values for this product for various types of rocks are listed in Table 5.

None of the rocks listed in Table 5, except the hematite ore, contained appreciable amounts of electronically conducting minerals. Rocks containing several percent of such minerals would have resistivity–dielectric constant products from 1 to 100 sec.

Average Electrical Properties of Large Masses of Rock

In Parkhomenko's monograph and in the preceding sections of this supplement, only the properties of single minerals and small samples of rocks have been considered. In many applications, it is necessary to consider the way in which the properties observed over a small volume of rock enter into an average for a larger volume of rock such as may determine the response to some surface-based exploration method. There are two types of gross character which must be considered in defining an average resistivity for a large volume of rock – that of a layered sequence of rocks, and that of faulted or jointed masses of dense rock. Only the layer case has been considered to date [71, 72, 116].

The layered model describes many of the rocks found at the earth's surface – sedimentary rocks, layered volcanic rocks, and igneous or metamorphic rocks which have been weathered to provide a zoned variation in electrical properties. The average electrical properties of a sequence of layered rocks may be described with a set of five parameters: ρ_1, the longitudinal resistivity, or average resistivity measured with current flowing parallel to the bedding planes; ρ_{tr}, the transverse resistivity, or the average resistivity measured with current flowing normal to the bedding plane; S, the intergrated conductivity for current flowing parallel to the bedding planes; T, the integrated resistivity for current flowing normal to the bedding planes; λ, the coefficient of anisotropy.

The five parameters are defined as follows:

$$T = \int_0^H \rho \, dh \tag{10}$$

$$S = \int_0^H \frac{1}{\rho}\ dh \qquad\qquad (11)$$

$$\rho_1 = H/S \qquad\qquad (12)$$

$$\rho_{tr} = T/H \qquad\qquad (13)$$

$$\lambda = (\rho_{tr}/\rho_1)^{\frac{1}{2}} \qquad\qquad (14)$$

In these equations, the resistivity is assumed to vary with depth, h, in a layer which has a total thickness, H.

The longitudinal resistivity, ρ_1, will always be less than the transverse resistivity, ρ_{tr}, unless the rock mass is completely uniform. This dependence of resistivity on the direction of current flow is anisotropy. A layered medium will appear to be anisotropic even though each layer in the medium is individually iso-

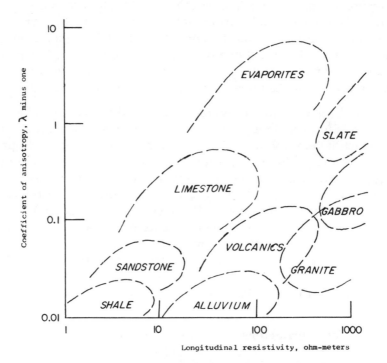

Fig. 8. Summary of common ranges for values for the coefficient of macroanisotropy and longitudinal resistivity for thick sequences of rock.

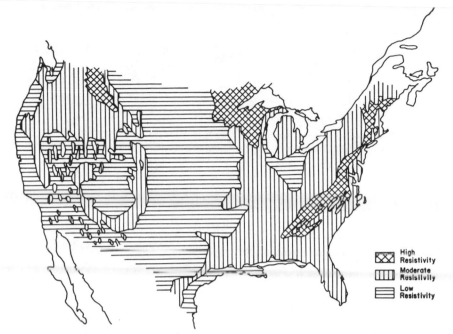

Fig. 9a. Ground resistivity map of the United States based on correlation between resistivities measured about radio stations and geologic sub-outcrop. From Keller and Frischknecht [75].

tropic. Therefore, this apparent anisotropy is called macroanisotropy or structural anisotropy.

Values for the average properties as defined here have been computed from electric logs in a number of areas [71, 72, 75], and appear to correlate consistently with both geological age and lithology as one might expect. The results of such complications can be summarized graphically as shown in Fig. 8.

Inasmuch as resistivity correlates with age and lithology, one might expect a map of the earth's resistivity to look much like a geological map. This is true if one generalizes resistivity data sufficiently so that the map is not confused by a wealth of local detail. Several resistivity maps of the United States based on resistivity determinations made at broadcast frequencies about radio stations have been published [32, 75]. The resistivity determined in such a manner represents an average value along a traverse from 5 to 30 miles long, but to depths of only several tens or hundreds of feet because of the limited penetration at radio frequencies.

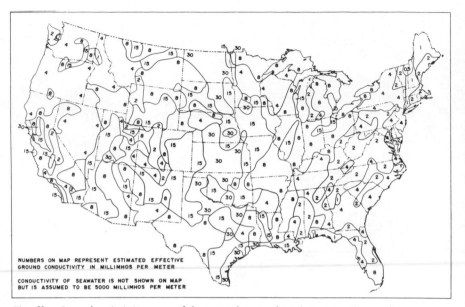

Fig. 9b. Ground resistivity map of the United States based on a correlation between resistivities measured about radio stations and soil type. From Fine [32].

Two such resistivity maps are shown in Fig. 9. The map in Fig. 9a is compiled on the basis that regions underlain by the same geological formation should have the same resistivity, and the boundaries of the various zones are drawn to coincide with geological boundaries [75]. The map in Fig. 9b is compiled on the basis that regions underlain by the same soil type should have the same resistivity, and the boundaries of the various zones are drawn to coincide with the boundaries between soil-type zones [32].

In spite of the differences, the two maps present much the same overall picture. High resistivity values are observed over areas underlain by crystalline rocks, such as the Appalachian area, the Canadian shield extension into the Wisconsin-Minnesota area, and in the Montana-Idaho area. Moderate resistivities are observed over areas underlain by older sedimentary rocks and by Tertiary volcanics, as in the southwestern United States. Low resistivities are observed over areas underlain by younger sedimentary rocks, particularly in the Great Plains area.

Bibliography

1.   L. A. Anderson. Electrical properties of sulfide ores in
      igneous and metamorphic rocks near East Union, Maine.
      In: Short Papers in the Geological Sciences, U. S. Geol.
      Surv. Profess. Papers 400B, pp. B125-B128, 1960.
2.   L. A. Anderson and G. V. Keller. A study in induced po-
      larization. Geophysics, Vol. 25, No. 5, pp. 848-864, 1964.
3.   G. E. Archie. The electrical resistivity log as an aid in
      determining some reservoir characteristics. Trans. AIME,
      Petrol. Br., Vol. 146, pp. 54-62, 1942.
4.   G. E. Archie. Electrical resistivity – an aid in core ana-
      lysis interpretation. Bull. Am. Assoc. Petrol. Geologists
      Vol. 31, No. 2, 1947.
5.   G. E. Archie. Classification of carbonate reservoir rocks
      and petrophysical considerations. Bull. Am. Assoc. Petrol.
      Geologists, Vol. 36, No. 2, 1951.
6.   L. F. Athy. Density, porosity, and compaction of sedimen-
      tary rocks. Bull. Am. Assoc. Petrol. Geologists, Vol. 14,
      pp. 1-24, 1930.
7.   M. L. Ayers, R. P. Dobyns, and R. Q. Bussell. Resis-
      tivities of water from subsurface formations. Petrol. Eng.,
      Vol. 24, No. 13, pp. B36-B48, 1952.
8.   H. Backström. Elektrisches u. thermisches Leitungsver-
      mogen des Eisenglanzes. Ofv. of Kongl. Vet.-Akad. Fork,
      Vol. 8, pp. 533-551, 1888.
9.   Carl Barus. On the electrical activity of ore bodies. In
      George F. Becker, 1882, Geology of the Comstock Lode and
      the Washoe District, U. S. Geol. Surv. Monograph 3, pp.
      309-367.
10.   H. Bloom. Molten electrolytes. In: Modern Aspects of
      Electrochemistry, No. 2, pp. 161-261, 1959.
11.   C. J. F. Böttcher. Theory of Dielectric Polarization.
      Elsevier Pub. Co., Amsterdam, 492 pp., 1952.
12.   W. F. Brace, A. S. Orange, and T. R. Madden. The effect
      of pressure on the electrical resistivity of water-saturated
      crystalline rock. J. Geophys. Res., Vol. 70, No. 22,
      pp. 5669-5698, 1965.
13.   R. S. Bradley, A. K. Jamil, and D. C. Munro. The elec-
      trical conductivity of olivene at high temperatures and pres-
      sures; Geochim. Cosmochim. Acta, Vol. 28, No. 11,
      pp. 1669-1678, 1964.

14. R. S. Bradley, A. K. Jamil, and D. C. Munro. The elec-
     trical conductivity of some organic and inorganic solids
     under pressure. II. Ferrous orthosilicates and spinel,
     Fe_2SiO_4. In: Symposium on the Physics and Chemistry of
     High Pressures, London, 1962, Gordon and Breach, Science
     Pub., Inc., pp. 147-149, 1963.

15. J. Brown. Artificial dielectrics. In: Progress in Dielec-
     trics, Vol. 2, J. Wiley and Sons, New York, pp. 193-225,
     1960.

16. J. A. V. Butler. Electrical Phenomena at Interfaces.
     MacMillan Co., New York, 309 pp., 1951.

17. L. C. Case. Subsurface water characteristics of Oklahoma
     and Kansas. In: Probl. of Petrol. Geol., Am. Assoc.
     Petrol. Geologists, 1934.

18. Hung-chi Chang, and George Jaffe. Polarization in electro-
     lytic solutions: Part I, Theory. J. Chem. Phys., Vol. 30,
     pp. 1071-1077, 1952.

19. I. I. Chebotarev. Metamorphism of natural waters in the
     crust of weathering. Geochim. Cosmochim. Acta, Vol. 8,
     pp. 52-63, 137-170, 1955.

20. L. G. Chombart. Well log interpretation in carbonate res-
     ervoirs. Geophysics, Vol. 25, pp. 779-853. 1960.

21. S. P. Clarke, Jr. (Ed.). Handbook of Physical Constants.
     Revised Edition, GSA Memoir 97, 587 pp., 1966.

22. R. H. Cole. Theories of dielectric polarization and relax-
     ation. In: Progress in Dielectrics, Vol. 3, pp. 47-100,
     J. Wiley and Sons, New York, 1961.

23. H. P. Coster. The electrical conductivity of rocks at high
     temperature. Roy. Astron. Soc. Monthly Notices, Geophys.
     Suppl., Vol. 5, pp. 193-199, 1948.

24. J. G. Crawford. Characteristics of oil-field waters of the
     Rocky Mountain region. In: Subsurface Geologic Methods,
     pp. 272-296, Colorado School of Mines, Golden, Colorado.

25. R. A. Daly, G. E. Manger, and S. P. Clarke, Jr. Density
     of rocks. In: Handbook of Physical Constants, Geol. Soc.
     Am., Mem. 97, pp. 19-26, 1966.

26. E. Darmois. Electrochimie. In: Handbuch der Physik,
     Vol. 20, Elektrische Leitungsphanomene II, pp. 392-479,
     Springer-Verlag, Berlin, 1957.

27. Denver Well Logging Society. Handbook of Rw values.
     Denver Well Logging Society, Denver, Colorado.

28. Michel Dumas. Resistivity and dielectric constant of rocks at sub-zero temperatures. M. Sc. Thesis, Colorado School of Mines, Golden, Colorado, 1962.

29. H. F. Dunlap, and R. R. Hawthorne. The calculation of water resistivity from chemical analyses. Trans. AIME, Vol. 192, pp. 373-375, 1951.

30. D. Eckhardt, K. Larner, and T. Madden. Long-period magnetic fluctuations and mantle electrical conductivity estimates. J. Geophys. Res., Vol. 68, No. 23, pp. 6279-6286, 1963.

31. I. Fatt. Effect of overburden and reservoir pressure on electrical logging formation factor. Bull. Am. Assoc. Petrol. Geologists, Vol. 41, No. 11, p. 3456, 1957.

32. H. Fine. An effective ground conductivity map for the continental United States. Proc. IRE, Vol. 42, pp. 1405-1408, 1954.

33. D. C. Fraser, N. B. Keevil, Jr., and S. H. Ward. Conductivity spectra of rocks from the Craigmont ore environment. Geophysics, Vol. 29, No. 5, pp. 832-847, 1965.

34. J. Frenkel. Kinetic Theory of Liquids. Dover Publ., Inc., New York, 488 pp. 1955.

35. Hugo Fricke. A mathematical treatment of the electrical conductivity and capacity of disperse systems. I: The electrical conductivity of a suspension of homogeneous spheroids. Phys. Rev., Vol. 24, pp. 575-587, 1923.

36. Hugo Fricke. A mathematical treatment of the electrical conductivity and capacity of disperse systems. II: The capacity of a suspension of conducting spheroids surrounded by a non-conducting membrane for a current of low frequency. Phys. Rev., Vol. 26, pp. 671-681, 1925.

37. Hugo Fricke. The electrical Permittivity of a dilute suspension of membrane-covered ellipsoids. J. Appl. Phys., Vol. 24, pp. 644-646, 1953.

38. R. H. Frische and H. von Buttlar. Induced electrical polarization. Geophysics, Vol. 22, No. 3, p. 688, 1957.

39. F. C. Frischknecht and L. A. Anderson. In situ measurements of the conductivity of molten lava in the Kilauea Iki crater, Hawaii. To be published, J. Geophys. Res., 1966.

40. F. C. Frischknecht and E. B. Ekren. Mapping conductive strata by electromagnetic methods. In: Short Papers in

Geological Sciences, U. S. Geol. Surv., Profess. Papers
400B, pp. B121-B125, 1960.

41.   H. Frölich. Theory of dielectrics. Oxford, Clarendon
      Press, 180 pp., 1949.

42.   G. F. J. Garlick. Photoconductivity. In: Handbuch der
      Physik, Vol. 19, Elektrische Leitungsphanomene I,
      Springer-Verlag, Berlin, pp. 316-395, 1956.

43.   A. N. Gerritsen. Metallic conductivity, experimental part.
      In: Handbuch der Physik, Vol. 19, Elektrische Leitungs-
      phanomene I, Springer-Verlag, Berlin, pp. 137-226, 1956.

44.   C. R. Glanville. Laboratory study indicates significant
      effects of pressure on resistivity of reservoir rock. J. Pet-
      rol. Technol., Vol. 11, No. 4, p. 20, 1959.

45.   Samuel Glasstone. An Introduction to Electrochemistry.
      D. Van Nostrand Co., New York, 1942.

46.   D. C. Grahame. Mathematical theory of faradaic admit-
      tance. J. Electrochem. Soc., Vol. 99, p. 370, 1952.

47.   R. M. Hamilton. Temperature variation at constant pres-
      sures of the electrical conductivity of periclase and olivene.
      J. Geophys. Res., Vol. 79, No. 22, pp. 5679-5692, 1965.

48.   R. D. Harvey. Electrical conductivity and polished mineral
      surfaces. Econ. Geol., Vol. 23, pp. 778-801, 1928.

49.   J. B. Hasted. The dielectric properties of water. In:
      Progress in Dielectrics, Vol. 3, J. Wiley and Sons, New
      York, pp. 101-150, 1961.

50.   D. P. Helander. The effect of pore configuration, pressure,
      and temperature on rock resistivity. In: Proc. SPWLA 7th
      Ann. Logging Symp., Soc. Profess. Well Log Analysts,
      Houston, Texas, 1966.

51.   D. P. Helander and J. M. Campbell. The effect of pore con-
      figuration, pressure, and temperature on rock resistivity.
      Ph. D. thesis, University of Oklahoma, 1965.

52.   J. D. Hem. Study and interpretation of the chemical ana-
      lyses of natural waters. U. S. Geol. Surv., Water Supply
      Papers 1473, GPO, Washington, D. C., 269 pp., 1959.

53.   D. W. Hilchie. The effect of pressure and temperature on
      the resistivity of rocks. Ph. D. thesis, University of
      Oklahoma, 1964.

54.   J. H. Hill, and J. D. Milburn. Effect of clay and water
      salinity on electrochemical behavior of reservoir rocks.
      AIME, J. Petrol. Technol., Vol. 207, p. 65, 1956.

55. Alan Holden. The Nature of Solids. Columbia University
     Press, New York, 1965.

56. D. S. Hughes and C. E. Cooke, Jr. The effect of pressure
     on the reduction of pore volume of consolidated sandstone.
     Geophysics, Vol. 28, No. 2, pp. 298-309, 1963.

57. Harry Hughes. The pressure effect on the electrical con-
     ductivity of peridote. J. Geophys. Res., Vol. 60, No. 2,
     pp. 187-191, 1955.

58. International Critical Tables, Vol. VI. McGraw-Hill Book
     Co., New York, p. 239, 1929.

59. A. F. Ioffe. Physics of Semiconductors. Academic Press,
     New York, 436 pp., 1960.

60. J. J. Jacobson. The electrical conductivity of olivene at
     high temperatures. M. Sc. thesis, Colorado School of
     Mines, Golden, Colorado, 1964.

61. George Jaffe and J. A. Rider. Polarization in electrolytic
     solutions. Part II: Measurements. J. Chem. Phys., Vol.
     20, pp. 1077-1087, 1952.

62. J. J. Jakosky, and R. H. Hopper. The effect of moisture
     on the direct-current resistivities of oil sands and rocks.
     Geophysics, Vol. 2, No. 1, pp. 33-35, 1937.

63. N. B. Keevil, Jr. and S. H. Ward. Electrolyte activity, its
     effect on induced polarization. Geophysics, Vol. 27, No. 5,
     pp. 677-690, 1962.

64. G. V. Keller. The effect of wettability on the electrical re-
     sistivity of sands. Oil Gas J., Vol. 51, pp. 62-65, 1953.

65. G. V. Keller. Electrical properties of sandstones of the
     Morrison formation. U. S. Geol. Surv. Bull. 1052J. pp.
     307-344, 1959.

66. G. V. Keller. Analysis of some electrical transient mea-
     surements on igneous, metamorphic and sedimentary rocks.
     In: Overvoltage Research and Geophysical Applications,
     Pergamon Press, London, pp. 92-111, 1959.

67. G. V. Keller. Physical properties of tuffs of the Oak Spring
     formation, Nevada. Short Papers in the Geological Sciences,
     U. S. Geol. Surv. Profess. Papers 400B, pp. B369-B400,
     1960.

68. G. V. Keller. Electrical properties of zinc-bearing rocks
     in Jefferson County, Tennessee. Short Papers in the Geo-
     logical Sciences, U. S. Geol. Surv. Profess. Papers.
     400B, pp. B128-B132, 1960.

69.   G. V. Keller. Pulse-transient behavior of brine-saturated
       sandstones. U. S. Geol. Surv. Bull. 1083D, pp. 111-129,
       1960.

70.   G. V. Keller. Electrical properties in the deep crust.
       IEEE Trans. Antennas Propagation, Vol. AP11, No. 3, pp.
       344-357, 1963.

71.   G. V. Keller. Compilation of electrical properties from
       electrical well logs. In: Quarterly, Colorado Schools of
       Mines, Vol. 59, No. 4, pp. 91-111, 1964.

72.   G. V. Keller. Statistical studies of electrical well logs.
       In: Trans. SPWLA 7th Ann. Well Logging Symp., Soc.
       Profess. Well Log Analysts, Houston, Texas, 1966.

73.   G. V. Keller. Electrical properties of rocks and minerals.
       In: Handbook of Physical Constants, Geol. Soc. Am., Mem.
       97, pp. 553-578, 1966.

74.   G. V. Keller. Geological survey investigations of the elec-
       trical properties of the crust and upper mantle. Geophysics,
       in print, 1966.

75.   G. V. Keller and F. C. Frischknecht. Electrical Methods
       in Geophysical Prospecting. Pergamon Press, London,
       526 pp., 1966.

76.   G. V. Keller and P. H. Licastro. Dielectric constant and re-
       sistivity of natural-state cores. U. S. Geol. Surv. Bull. 1052H.
       pp. 257-286, 1959.

77.   G. V. Keller, J. I. Pritchard, L. A. Anderson, and
       C. J. Zablocki. Electrical surveys on the Island of Hawaii.
       To be published, J. Geophys. Res., 1966.

78.   Charles Kittel. Introduction to Solid State Physics. J. Wiley
       and Sons, New York, 1953.

79.   B. N. Lahiri and A. T. Price. Electromagnetic induction
       in non-uniform conductors, and the determination of the
       conductivity of the earth from terrestrial magnetic varia-
       tions. Roy. Soc. (London), Phil. Trans., Vol. 237A,
       pp. 509-540, 1939.

80.   W. H. K. Lee (Ed.). Terrestrial heat flow. Geophys.
       Monograph Ser. 8, Am. Geophys. Union, Washington, D. C.,
       267 pp., 1965.

81.   P. H. Licastro and G. V. Keller. Resistivity measure-
       ments as a criterion for determining fluid distributions in
       the Bradford sand. Producers Monthly, Vol. 17, No. 7,
       pp. 17-23, 1953.

82. A. B. Lidiard. Ionic Conductivity. In: Handbuch der Physik, Vol. 20, Elektrische Leitungsphanomene II, Springer-Verlag, Berlin, pp. 246-349, 1957.

83. J. R. Macdonald. Dielectric dispersion in materials having a distribution of relaxation times. J. Chem. Phys., Vol. 20, pp, 1107-1111, 1952.

84. Otfried Madelung. Halbleiter. In: Handbuch der Physick, Vol. 20, Elektrische Leitungsphanomene II, Springer-Verlag, Berlin, pp. 1-245, 1957.

85. G. E. Manger. Porosity and bulk density of sedimentary rocks. U. S. Geol. Surv. Bull. 1144E, pp. E1-E55, 1963.

86. D. J. Marshall and T. R. Madden. Induced polarization – a study of its causes. Geophysics, Vol. 24, No. 4, pp. 790-816, 1959.

87. J. C. Maxwell. Theory of a composite dielectric. In: Treatise on Electricity and Magnetism, Dover Publ., New York, 1891.

88. K. L. McDonald. Penetration of the geomagnetic secular field through a mantle of variable conductivity. J. Geophys. Res., Vol. 62, No. 1, pp. 117-141, 1957.

89. J. G. McKelvey, Jr., P. F. Southwick, K. S. Spiegler, and M. R. J. Wyllie. The application of a three-element model to the S. P. and resistivity phenomena evinced by dirty sand. Geophysics, Vol. 20, pp. 913-931, 1955.

90. R. J. Meakins. Mechanisms for dielectric adsorption in solids. In: Progress in Diclcctrics, Vol. 3, pp. 151-202, J. Wiley and Sons, New York, 1961.

91. O. E. Meinzer (Ed.). Hydrology. Dover Publ., New York, 712 pp., 1942.

92. L. Migeaux, J.-L. Astier, and P. Revol. Essai de determination experimental de la resistivite electrique des couches profondes de l'ecorce terrestre. Compt. Rend. Acad. Sci., Vol. 251, pp. 567-569, 1960.

93. K. R. Miles. Origin and salinity distribution of artesian water in the Adelaide Plains, South Australia. Econ. Geol., Vol. 46, pp. 193-207, 1951.

94. E. J. Moore. A graphical description of a new method for determining equivalent NaCl concentration from chemical analyses. In: Trans. SPWLA 7th Ann. Well Logging Symp., Soc. Profess. Well Log Analysts, Houston, Texas, 1966.

95. E. J. Moore, S. E. Szasz, and B. F. Whitney. Determining formation water resistivity from chemical analyses. J. Petrol. Technol., March, 1966.

96. N. F. Mott and H. Jones. The Theory of the Properties of Metals and Alloys. Dover Publ., New York, 326 pp., 1958.

97. K. Noritomi. Studies on the change of electrical conductivity with temperature of a few silicate minerals. Tohoku Univ. Sci. Rept., 5th Ser., Vol. 7, No. 3, pp. 119-125, 1954.

98. K. Noritomi. Migration of charge carriers in the case of electrical conductivity of rocks; Tohoku Univ. Sci. Rept., 5th Ser., Vol. 9, No. 9, pp. 120-127, 1958.

99. K. Noritomi. The electrical conductivity of rock and the determination of the electrical conductivity of the earth's interior. J. Mining Coll., Akita Univ., Ser. A., Vol. 1, No. 1, pp. 27-59.

100. K. Noritomi and A. Asada. Studies on the electrical conductivity of a few samples of granite and andesite. Tohoku Univ. Sci. Rept., 5th Ser., Vol. 7, No. 3, pp. 201-207, 1957.

101. D. S. Parasnis. The electrical resistivity of some sulfide and oxide minerals and their ores. Geophys. Prospecting, Vol. 4, pp. 249-278, 1956.

102. H. W. Patnode and M. R. J. Wyllie. The presence of conductive solids in reservoir rock as a factor in electric log interpretations. Trans. AIME, Vol. 189, p. 47, 1950.

103. L. Pauling. The Nature of the Chemical Bond. Cornell University Press, Ithaca, New York, 644 pp., 1960.

104. F. M. Perkins, Jr., H. R. Brannon, and W. O. Winsauer. Interrelation of resistivity and potential of shaly reservoir rock. Trans. AIME, Vol. 201, p. 176, 1954.

105. S. J. Pirson. Handbook of Well Log Analysis. Prentice-Hall, Englewood Cliffs, New Jersey, 326 pp., 1963.

106. K. W. Plessner and R. West. High-permittivity ceramics for capacitors. In: Progress in Dielectrics, Vol. 2, J. Wiley and Sons, New York, pp. 165-192, 1960.

107. P. Popper. Non-oxide ceramic dielectrics. In: Progress in Dielectrics, Vol. 1, J. Wiley and Sons, New York, pp. 217-270, 1959.

108. A. Poupon, M. E. Loy, and M. P. Tixier. A contribution to electric log interpretation in shaly sands. Trans. AIME, Vol. 201, pp. 138-145, 1954.

109. L. A. Puzin. Connate water resistivity in Oklahoma – its application to electric log interpretation. Petrol. Eng., Vol. 24, No. 9, pp. B67-B77, 1952.

110. J. C. Redmond. Effect of simulated overburden pressure on resistivity, porosity, and permeability of selected sandstones. Ph. D. Thesis, Pennsylvania State University, 1962.

111. T. Rikitake. Electromagnetism and the Earth's Interior. Elsevier Publ. Co., Amsterdam, 307 pp., 1966.

112. S. K. Runcorn and D. C. Tozer. The electrical conductivity of olivene at high temperature and pressure. Ann. Geophys., Vol. 11, No. 1, pp. 98-102, 1955.

113. J. F. Sage. Water analysis. In: Subsurface Geology in Petroleum Exploration, Colorado School of Mines, Golden, Colorado, pp. 251-264, 1958.

114. V. V. J. Sarma and V. B. Rao. Variation of electrical resistivity of river sands, calcite, and quartz powders with water content. Geophysics, Vol. 27, No. 4, pp. 470-479, 1962.

115. J. A. Schufle. Cation exchange and induced electrical polarization. Geophysics, Vol. 24, No. 1, p. 164, 1959.

116. C. Schlumberger, M. Schlumberger, and E. G. Leonarden. Some observations concerning electrical measurements in anisotropic media and their interpretation. Trans. AIME, Vol. 110, p. 159, 1934.

117. J. H. Scott. Electric and magnetic properties of rock and soil. Unpublished report, U. S. Geol. Surv., Denver, Colorado.

118. J. C. Slater. The electronic structure of solids. In: Handbuch der Physik, Vol. 19, Elektrische Leitungsphanomene I, Springer-Verlag, Berlin. pp. 1-136, 1956.

119. Smythe and Hitchcock. Electrical properties of ice. J. Am. Chem. Soc., Vol. 54, p. 4631, 1932.

120. J. M. Stevels. The electrical properties of glass. In: Handbuch der Physik, Vol. 20, Elektrische Leitungspanomene II, Springer-Verlag, Berlin, pp. 350-391, 1957.

121. V. T. Stringfield, M. A. Warren, and H. H. Copper, Jr. Artesian water in the coastal area of Georgia and Florida. Econ. Geol., Vol. 36, pp. 698-711, 1941.

122. K. Sundberg. Geophysical prospecting. AIME, p. 381, 1932.

123. C. Suskind. On the isotropic artificial dielectric. Proc. IRE, Vol. 40, p. 1251, 1952.
124. P. M. Sutton. The dielectric properties of glass. In: Progress in Dielectrics, Vol. 2, J. Wiley and Sons, New York, pp. 113-164, 1960.
125. S. A. Sweeney and H. Y. Jennings. The electrical resistivity of preferentially water wet and preferentially oil wet carbonate rocks. Producers Monthly, p. 29, May, 1960.
126. G. Towle. An analysis of the formation resistivity factor–porosity relationship of some assumed pore geometries. In: Trans. SPWLA 3rd Ann. Well Logging Symp., Soc. Profess. Well Log Analysts, Houston, Texas, 1962.
127. D. C. Tozer. The electrical properties of the earth's interior. In: Physics and Chemistry of the Earth. VIII. Pergamon Press, New York, pp. 414-436, 1959.
128. A. R. von Hippel. Dielectrics and Waves. J. Wiley and Sons, New York, 284 pp., 1954.
129. A. R. von Hippel. Dielectric Materials and Applications. J. Wiley and Sons, New York, 1954.
130. K. W. Wagner. Zur theorie der Unvolkommenen Dielektrik. Ann. der Phyz., Vol. 40, pp. 817-855, 1913.
131. J. R. Wait. Discussions on "A theoretical study of induced electrical polarization". Geophysics, Vol. 23, No. 1, pp. 144-152, 1958.
132. J. R. Wait (Ed.). Overvoltage Research and Geophysical Applications. Pergamon Press, London, 158 pp., 1959.
133. W. O. Winsauer and W. M. McCardell. Resistivity of brine saturated sands in relation to pore geometry. Trans. AIME, Vol. 198, p. 129, 1953.
134. W. O. Winsauer et al. Resistivity of brine-saturated sands in relation to pore geometry. Bull. Am. Assoc. Petrol. Geologists Vol. 36, p. 278, 1952.
135. W. O. Winsauer, F. M. Perkins, Jr., and H. R. Brannon, Jr. Interrelation of resistivity and potential of shaly reservoir rock. J. Petrol. Technol., Vol. 6, No. 8, pp. 29-34, 1954.
136. D. O. Wyble. Effect of applied pressure on conductivity and permeability of sandstones. J. Petrol. Technol., Vol. 10, p. 57, 1959.

137. G. Wyllie. Theory of polarization and absorption in dielectrics – an introductory survey. In: Progress in Dielectrics, Vol. 2, J. Wiley and Sons, New York, pp. 1-28, 1960.

138. M. R. J. Wyllie. Log interpretation in sandstone reservoirs. Geophysics, Vol. 25, No. 4, pp. 748-778, 1960.

139. M. R. J. Wyllie and A. R. Gregory. Formation factor of unconsolidated porous media: influence of particle shape and effect of cementation. Trans. AIME, Vol. 198, p. 103, 1953.

140. M. R. J. Wyllie and W. D.Rose. Some theoretical considerations related to the quantitative evaluations of the physical characteristics of reservoir rock from electric log data. Trans. AIME, Vol. 189, p. 105, 1950.

141. M. R. J. Wyllie and M. B. Spangler. Application of electrical resistivity measurements to the problem of fluid flow in porous media. Bull. Am. Assoc. Petrol. Geologist, Vol. 36, No. 2, 1951.

142. W. A. Yager. The distribution of relaxation times in typical dielectrics. Phys., Vol. 7, pp. 434-450, 1936.

143. C. J. Zablocki. Measurement of the electrical properties of rocks in southeast Missouri. In: Short Papers in the Geological and Hydrologic Sciences, U. S. Geol. Surv. Profess. Papers 400B, pp. B214-216, 1960.

144. C. J. Zablocki. Electrical properties of sulfide-mineralized gabbro, St. Louis County, Minnesota. In: Short Papers in the Geological and Hydrologic Sciences; U. S. Geol. Surv. Profess. Papers.424C, pp. C256-C258, 1961.

145. C. J. Zablocki. Electrical and magnetic properties of a replacement type magnetite deposite in San Bernadino County, California. In: Short Papers in Geology, Hydrology, and Topography, U. S. Geol. Surv. Profess. Papers 450D, pp. D103-D104, 1962.

146. C. J. Zablocki. Electrical properties of serpentinite from Mayaguez, Puerto Rico. In: A Study of Serpentinite, Natl. Acad. Sci. Natl. Res. Council, Publ. No. 1188, pp. 107-117, Washington, D. C., 1964.

147. C. J. Zablocki. Some applications of geophysical logging methods in mining exploration drill holes. In: Trans. SPWLA 7th Ann. Well Logging Symp., Soc. Profess. Well Log Analysts, Houston, Texas, 1966.

148. C. J. Zablocki and G. V. Keller. Borehole geophysical
      logging methods in the Lake Superior district. In: Proc.
      7th Ann. Drilling Expl. Symp., University of Minesota,
      Minneapolis, 1957.

INDEX

Index

311

Dielectric loss (continued)
  effect of temperature and pressure, 200-215
  for laminated material, 191
Dielectric loss tangent, 187
  for minerals, 217
Dielectric material, 59, 94
  conduction, 60-63
  ionic conduction, 62
  surface conduction, 73
Dielectric polarization, 13
Dipole relaxation, 16
Donor, 77

Elastic properties, correlation with electrical properties, 138
Electrochemical processes at electrodes, 218
Electrolytes, solid, 59, 266
Essential minerals, 2

Falkenhagen formula, 44
Faraday's law, 61, 62
Formation factor, 124-125, 127, 139

Graphite, 91
Ground water 271, 272
  geochemical classification, 272
  resistivity, 273

Halides, dielectric constant, 24, 30
Hall effect, 61-62, 74, 91, 92
"Hole" conduction, 76-77
Hydroxides, dielectric constant, 26, 30

Ice, dielectric dispersion, 235, 236
Igneous rocks, 3
  dielectric constant, 39
  dielectric dispersion, 230-236
Induced polarization, 286
Ionic mobility, function of temperature, 64

Joints
  classification, 136-137
  effect on anisotropy, 136

Lichteniker's mixing rule, 20
Lorentz-Lorenz mixing rule, 20

Madelung constant, 53
Maxwell's mixing rule, 20
Maxwell-Wagner theory, 221
Metamorphic rocks, 3
  dielectric constant, 38
  dielectric dispersion, 230-236
Methods of measurement
  bridge methods, 197
  dielectric constant and loss, 195-200
  effect on results, 86
  four-electrode method, 83
  frequency ranges, 200
  resistivity, 79-85
  resistivity under pressure, 161
  thermal loss method, 219
  two-electrode method, 80
Mica, dielectric properties, 201
Molten rocks, resistivity, 286
Molybdenates, dielectric constant, 27
Monomineralic rocks, 2

n-type conduction, 77

Odelevskii's mixing rule, 20
Ohm's law, 73
Oil
  effect on resistivity, 129
  in hydrophilic rocks, 131-132
  in hydrophobic rocks, 131-132
Onsager relation, 17
Ores
  copper-nickel, 105
  copper-sulfur, 101
  iron, 103, 105
  polymetallic sulfide, 102
  pyrite, 100
  resistivity, 88, 90, 108-109, 110
  zinc, 104
Overvoltage, 290
Oxides, dielectric constant, 25, 30-31

Permeability, dielectric, 11
  of carbonates, 26
  of elements, 23, 24
  of halides, 24
  of hydroxides, 26
  for metal sulfides, 23, 24
  of mixtures, 19